国际展会实务（双语版）

International Exhibition Practice（Bilingual Edition）

主　编	崔翔勇		
副主编	董　莉	崔苪萱	贺宇涛
	王继旭	陈金丹	闫欣瑞
编　者	田倩倩	成新霞	马欢欢
	姚　萍	田文菡	

北京理工大学出版社
BEIJING INSTITUTE OF TECHNOLOGY PRESS

内 容 简 介

本书是探索校企合作人才培养模式的成果，内容涉及展会的作用、展位分类、设计与搭建、展馆规则、展前、展中、展后的相关筹备工作、展会预算、潜在客户转化、展品运输、全球主要展会业发展概况、知识产权保护、国际会展城市和人文地理等。

本书主编在国际展会领域工作了近 30 年，有丰富的实战经验，在编写过程中突出实务，淡化理论，将国际展会各链条知识有机串联在一起。

本书实用性强，以提升学生的综合素养为主要目标，可作为商务英语、会展经济与管理、国际经济与贸易等本专科学生的教材，也可以作为外贸企业的培训教材。

图书在版编目（CIP）数据

国际展会实务：双语版／崔翔勇主编. --北京：
北京理工大学出版社，2023.6
ISBN 978-7-5763-2501-0

Ⅰ.①国… Ⅱ.①崔… Ⅲ.①展览会-双语教学-高
等学校-教材 Ⅳ.①G245

中国国家版本馆 CIP 数据核字（2023）第 113365 号

出版发行／北京理工大学出版社有限责任公司

社　　址／北京市海淀区中关村南大街 5 号

邮　　编／100081

电　　话／（010）68914775（总编室）
　　　　　（010）82562903（教材售后服务热线）
　　　　　（010）68944723（其他图书服务热线）

网　　址／http://www.bitpress.com.cn

经　　销／全国各地新华书店

印　　刷／三河市天利华印刷装订有限公司

开　　本／787 毫米×1092 毫米　1/16

印　　张／14.75　　　　　　　　　　　　责任编辑／王梦春

字　　数／344 千字　　　　　　　　　　　文案编辑／杜　枝

版　　次／2023 年 6 月第 1 版　2023 年 6 月第 1 次印刷　　责任校对／刘亚男

定　　价／89.00 元　　　　　　　　　　　责任印制／李志强

序

提笔写下这些文字的时候，我脑海里浮现出当年我和崔翔勇老师见面的场景——在北京国贸酒店大堂的咖啡吧。岁月如梭，转眼已经与崔老师相识多年，时光的刻刀并没有在他脸上留下太多的雕琢，也丝毫没有浇灭他对会展行业的满腔热忱。

展览人写展览，而且能沉淀下来编写成书的人真的少之又少，同为展览人我深感欣慰。正所谓纸上谈来终觉浅，在这本书中，涵盖的并不仅仅是丰富且专业的理论，更是作者从多年工作历练中提炼出来的宝贵经验，具有很强的实战性。

会展业是德国经济的重要组成部分，根据德国展览业协会（AUMA）2018 年发布的"德国展览业整体经济相关性"报告（由伊福经济研究所霍斯特·彭茨科弗先生代表 AUMA 所作），来自全球各地的参展商和参观者在 2014—2017 年度，平均每年在德国的支出达 145 亿欧元，创造了 23.1 万个工作岗位。

这几年，疫情对德国会展行业造成了较大冲击。从 2020 年 3 月到 2021 年 9 月，德国贸易展览会举办地被迫停滞了近 19 个月。2020 年，仅对展会组织者而言，就导致了 70% 的销售损失。到目前为止，总的经济损失已达 420 亿欧元。2021 年，又一次出现巨大损失。截至 2021 年 9 月底，预定举办的 380 场贸易展会中有三分之二以上不得不取消。与此同时，德国会展行业在积极探索新的办会模式。德国展览业协会表示，展会组织者加速数字化转型，以维持与客户的联系。据德国会议促进局（GCB）2022 年 5 月公布的《会议和活动晴雨表》，2021 年，混合模式会议活动数量增加了 280%，线上虚拟会议活动数量增加了 120%。在参与人数方面，2021 年，参加混合模式会议活动的人数为 1 840 万，而在 2020 年只有 180 万人。

众所周知，德国是世界排名第一的国际贸易展览会举办地，近三分之二的不同行业领域的全球领先贸易展览会在德国举行。我所在的德国科隆展览有限公司是世界领先的贸易展览会、会议和活动组织者之一，拥有自己的展览场地。"我们已万事俱备，只待你们重整旗鼓！"这是我们在 2020 年写下的结束语。机会只留给有准备的人，作为展会主办方我们随时都准备好了，携手我们遍布全球的贸易展会的客户，重启的共同愿景在 2021 年秋季实现了。

到 2034 年前，"科隆展览 3.0"计划将投入总额超过 7 亿欧元用于提升和发展贸易展览场地，以使其能够覆盖未来所有的活动形式。这将是科隆展览历史上最广泛的投资计划。而作为该计划的重要组成部分，集贸易展会、会议和活动等功能于一身的 Confex 会展综合体，已于 2022 年 11 月 21 日成功举行了封顶仪式，并将于 2024 年开始投入使用。由 UFI-全球展览业协会主办的展览行业年度全球大会目前已经确定 2024 年在 Confex 的会议中心举办。毋庸置疑，科隆将成为比以往任何时候都更具吸引力的会议胜地，这也将是科隆展览成立 100 周年的一大亮点。

我们的会展场馆在德国排名第三并进入全球前十名，每年在科隆和世界各地的主要市场举办约 80 个交易会、展览、客座展览和特别活动。11 个展馆，近 40 万平方米的展区和室外区域，17 500 个停车位，每年 2 000 场会议，接待来自 122 个国家的约 54 500 家参展公司，来自 224 个国家的约 300 万观众，约 22 万名注册的记者，超过 10 亿个广告和媒体联络人，我们在 100 多个国家设有海外代表处。

我们以无限的热情和丰富的经验，活跃于科隆和全球的国际贸易展览会业务。从家具到食品，从五金到户外，从移动出行到游戏，通过我们的活动，我们达到了最高质量标准，并在工业、商业、政治、服务和消费品领域之间建立了全球接口。

国际展览会从业者非常辛苦，要常年在全球各地飞来飞去，但同时也令人羡慕，因为可以在工作之余体验世界各地风土人情，开阔眼界；更重要的是，我们可以在第一时间了解到每个行业最新的发展动态，因为展览会是全球贸易交易的最前沿阵地，也是学习产品知识、结识客户和了解市场的生动课堂。

崔翔勇老师就是这样一位国际展会的从业者，2019 年他去过了全球 12 个国家的 26 个城市，飞行里程超过了 20 万千米，是我们展览界的"飞人"。正是他的这种经历，让他有机会对会展相关的知识进行积累和沉淀，并能在适当的时候把这些知识整理成书，以教材的方式分享给年轻的一代。

同为展览人，我们一生中有许多回忆与故事都是围绕着展会展开的，或在星夜赶往展会的航班上，或从展馆返回酒店的大巴车上，或在银装素裹的寒冬，或在骄阳似火的盛夏，或在展馆附近的中餐馆享受美食的片刻，或在候机厅内凝视着落日的一瞬间。能把工作经历，把自己的苦与乐灌注在笔尖，是展览人在"孤寂"日子里对所知所获的分享。在《国际展会实务》这本书里，我看到了自己与自己心灵的对话，不需要解说，只要自己懂得就是。

有位作家说过，日子长了，不知不觉就攒下了这么些叫作文章的东西。我想，这就是崔翔勇老师写下了这些文字的缘故。

崔国豪

科隆展览（北京）有限公司董事总经理

前　言

　　二十大报告提出要加强基础学科、新兴学科、交叉学科建设，加快建设中国特色、世界一流的大学和优势学科。敏锐地关注到当代学科发展的趋势及其对于科技发展与一流大学建设的作用。报告还提出要加强教材建设和管理，学科建设和教材建设问题，也是第一次出现在党代会的报告之中。

　　二十大报告提出：要"构建优质高效的服务业新体系，推动现代服务业同先进制造业、现代农业深度融合"。会展业作为现代服务业的重要组成部分，是经济的"助推器"，也是衡量区域开放度、经济活力和发展潜力的重要标志之一。

　　习近平总书记在十四届全国人大一次会议闭幕会上指出，"中国的发展惠及世界，中国的发展离不开世界。我们要扎实推进高水平对外开放，既用好全球市场和资源发展自己，又推动世界共同发展。"

　　推动贸易创新发展、加快建设贸易强国是我国推进高水平对外开放的重要组成部分，也是更好地畅通国际大循环、与世界共同发展的题中应有之义。外贸进出口是拉动经济增长的重要引擎。过去 5 年，我国坚定扩大对外开放，推动外贸进出口稳中提质。货物进出口总额年均增长 8.6%，突破 40 万亿元，连续多年居世界首位。

　　国际贸易展览会（也可称为展会）是全球商品和服务贸易的主要推动力，是 B2B 的重要营销平台。国际性展会加剧了全球几乎所有领域的竞争和贸易往来，确保了经济增长和就业。经济全球化和对品牌的日益关注是促进会展业在全球范围内迅速发展的另一重要因素。在此过程中，展会组织者正在以更全面的方式发展成为企业的营销合作伙伴。此外，展览会对推动国家经济建设发展和促进国民经济效益增长发挥着日益重要的作用，得到了社会各界的高度重视。

　　目前，会展业已发展成为新兴的现代服务型产业，成为衡量一个城市国际化程度和经济发展水平的重要标准之一。根据全球展览业协会（Union of International Fairs，UFI）的数据，全球每年在约 1 200 个展览场馆举办 31 000 场展览会，展览面积超过 3 480 万平方米，相当于 460 个足球场。每年约有 440 万名参展商和超过 2.6 亿观众聚集在展会上，可提供 231 000 个工作岗位。

　　据报道，德国国家旅游局、欧洲活动中心协会和德国会议促进局联合发布报告，对德国 2021 年会展市场的关键数据进行了分析，认为会展行业在新冠肺炎疫情影响下正积极谋求转型，线上虚拟会展、线上线下相结合的混合会展以及可持续性会展将成新趋势。

在国际性的展会上，主办方、参展商和采购商在大部分情况下都是以英语作为工作语言，涉及大量的专业词汇和术语。为适应国际展览行业人才培养需要，本书将理论与实践相结合，构建以展览会流程为导向的教学内容体系，融合多学科知识、语言沟通技巧，力求把学生培养成为能满足社会需求的复合型、应用型人才。

本书的主编崔翔勇老师在国际展会领域工作了近 30 年，曾赴 60 个国家和地区参加500 场次的展览会，具有丰富的实战经验。因此本书在编写过程中突出了实务，淡化了理论，以国际展会为背景，把全行业各链条的知识有机地串联在一起，给读者耳目一新的感受。

本书是校企合作培养模式的成果，编者根据会展行业的工作环境和岗位要求，营造了仿真的工作情境，使学生在逐项了解展会工作任务的同时，能掌握相应的英语词汇、表达方法以及国际贸易中涉及的展会选择、产品营销、客户管理及沟通技巧。在课程完成后，学生可以掌握国际展览活动中的基本知识，提升专业英语运用能力，培养通过国际展会开拓市场的技能。

本书共设 10 个单元，每个单元均设有本章导读、知识精讲、展会知识并包括 3 篇阅读文章。本章导读汇总了各章节的知识要点，展会知识是每节课的必读材料，涵盖了国际展会的多重知识，包括全球会展业发展概况、新冠疫情对全球会展行业的影响、后疫情时代国际会展的变化、一些重要国际展会的介绍、世博会相关知识、知识产权保护等。强烈建议教师能带领学生认真研读此部分的内容，并对相关知识进行汇总，这有助于学生积累国际展会知识与信息。

本书的正文部分涉及国际展会的很多知识要点：展会的作用，展位分类、设计与搭建，展馆规则，展前、展中、展后的相关筹备工作，展会预算，潜在客户转化，展品运输，全球主要会展业发展概况，知识产权保护，国际展会城市和人文地理等。

每个单元的第 1 篇文章为精读材料，后面配有练习题，有助于总结主要知识点；第 2篇和第 3 篇文章为泛读材料，但建议学生在教师的指导下认真通读，并以小组讨论的方式对文中有用的知识点进行讨论，任课教师在教学过程中也可根据课时灵活掌握。本书的练习以提高学生的专业知识能力和英语语言能力为出发点，分为简答题、选择题、翻译题、英文写作几部分。本书还在每个单元最后增加了常用展会及工作情境对话练习。

本书主要有以下特色：

1. 创新性强

本书主编是工作在国际展会一线的专家，对国际性展会非常了解，深知从事这个行业应该掌握的知识，因此对于本书的内容进行了精挑细选，教材内容具有典型性、真实性及前沿性的特点，对未来从事国际展会和国际贸易销售岗位的人员具有重要的指导意义。本书创新性地设计了知识精讲和展会知识两个栏目，既能满足学生个人对短视频教学的需求，又能通过阅读掌握国际展会相关知识，特别是讨论和小组分享等能培养学生团队项目合作精神，树立展会职业道德意识。

2. 专业性强

本书用英语介绍了大量国际著名展会，涉及全球重要展会活动，还涉及一些组展和参展的经验。本书的难易程度为中等，可以作为商务英语、展会专业和国际贸易专业学生的教材，也可以作为专业培训班的教材。

在素材整理过程中，我们注重融教材的专业性、实践性、趣味性于一体，以提高学生解决实际问题的能力。

3. 实操性强

本书以展会企业和国际贸易企业外销员岗位需求为出发点，以国际性展览商务活动体系中的典型工作任务为导向，从主办方、参展商，以及采购商三种不同角色的真实工作任务出发，编排教学内容，主要内容涵盖了国际展会的基本流程，还涉及与展会相关的行业与话题，如展会城市、知识产权、国际贸易、商务谈判等。

在本书的编写和策划过程中，我们得到了河北科技大学、石家庄学院、河北工业职业技术大学、河北政法职业学院的支持，同时河北爱德会展服务有限公司、思鸽供应链管理河北有限公司、河北昭阳货运代理有限公司、石家庄市进出口企业协会、河北创富蓝海国际贸易咨询服务有限公司、石家庄德卡诺装饰材料有限公司、河北华星尔贸易有限公司、石家庄万维贸易有限公司、河北福罗德地板材料科技有限公司、邢台三厦铸铁有限公司、石家庄金萱商贸有限公司、石家庄傲瑟进出口贸易有限公司、石家庄华莹玻璃制品有限公司、河北越洋国际货运代理有限公司等企业对本书的编写提出了非常好的建议。

本书由崔翔勇担任主编，统筹全书内容，并对所有素材进行了审定，河北科技大学董莉负责编写第一章，美国 Sam Houston State University Dixuan Cui（崔苮萱）负责编写第十章，石家庄学院贺宇涛负责编写第二章，哈尔滨剑桥学院陈金丹负责编写第三章，山西工商学院闫欣瑞负责编写第四章，河北科技大学田倩倩负责编写第五章，石家庄学院成新霞负责编写第六章，河北政法职业学院马欢欢负责编写第九章，河北工业职业技术大学姚萍负责编写第八章，石家庄学院田文菡负责编写第七章，石家庄德卡诺装饰材料有限公司董事长王继旭为本书编写提供了诸多案例及行业信息，在此对作者们的辛勤付出致以衷心的感谢。

尽管本书编者在特色建设方面作出了许多努力，但由于水平所限，难免有疏漏之处，恳请广大读者批评、指正。

编　者

目　录

UNIT 1

TRADE FAIRS

本章导读

通过本章的学习，了解展览会的作用，特别是其在国际贸易中的作用。国际性展览会是全球贸易合作交流的平台，是产品、想法和技术的聚集场所和配送中心。企业在展会上可以实现全方位的营销目标。一场展览会通常只有3~5天的时间，在短短几天内，不同的产品和服务就可受到市场的考验。

深入理解展览会在开拓国际市场方面的重要作用。参加一场国际展会可以分析出一个公司的竞争优势和劣势。这种竞争分析可以让一家企业更明智地评估自己的市场地位，学习竞争对手的市场行为，更好地评估自己的发展。

国际展览会的作用有哪些？展览会是促进买卖双方成交的场合，是发布新产品和新信息的舞台，是企业学习各国先进技术和产品设计的课堂，是同业之间交流的平台，是企业进行广告宣传的途径。

知识精讲

展会知识

对期待复苏、回归展会现场重建商业连接的行业提供驱动和支持

通常被称为贸易展览或交易会的展览有一个简单的目的——将各行业聚集在一起，建立行业社群，创造供应链机会。但这个行业本身并不简单：对与会者而言，展会促成超过493亿欧元（551亿美元）的商业销售额，对全球经济产生了难以置信的影响。

此外，观众和参展商产生的相关参观参展费用总计达299亿欧元（334亿美元），对会展业价值链（组织者、场馆和服务提供商）和相关旅游活动（住宿、餐饮和旅游）也影响重大。

在一场使商业和区域经济都濒临崩溃的重大疫情之后，贸易平台为重建在整个疫情期间相对停滞的供应链提供了机遇，使买卖双方在最需要的时候建立网络、开展业务、学习知识和创新产品。

贸易展览会历经多年发展，如今已远不只是展示产品的展位。在将富有价值的数字技术进步与当下令人渴望的人际关系结合起来的社交网络活动的支持下，展览会已经成为由

教育机会、创新发布、数据驱动的符合条件的潜在客户信息获取机会等多种功能所增强的行业市场。2022 年，在全球举办的 33 000 个展会中，许多都转向了数字解决方案，这些解决方案利用了一流的技术来创建关键的供应链连接。

尽管数字解决方案在新冠疫情期间被证明是非常有价值的，但展会依然随着各经济体针对疫情控制的有利进展而重新开放，调研数据显示供应链的双方都渴望回到面对面平台。

Informa Markets 是全球最大的展会组织者之一，自 2020 年 6 月以来，其在中国举办的贸易展开始强劲复苏。CBME 作为母婴行业的领先平台吸引了超过 9 万名观众，M-cro 贸易（上海）副总裁 Julian van Gemeren 说，这是一个富有成效的展会。他证实："我们拓展了分销渠道，会见了许多客户，仅第一天就收获了 70 家新经销商。"

B2B 活动组织者 Tarsus Group 也见证了现场展会活动的强劲回归，最近其在中国举办的展会的参展人数甚至超过了疫情之前的水平。该主办方还恢复了在美国和墨西哥等地区的展会活动，其中 Expo Manufactura 的参与者表达了对面对面活动回归的欢迎和渴望。

Bricos 的 Emilio Cuervo 评论说："参加这个展会非常重要，因为它使我们能够直接接触我们的现有客户和潜在客户。2021 年没有举办这项活动的影响对我们来说显而易见，因为我们的一些客户和销售线索都是由 Expo Manufactura 给予的。"

北美最大的活动策划商之一——Emerald 在佛罗里达州的奥兰治县会议中心成功举办了美国 2022 年第一个现场活动 Surf Expo Winter。客户的积极情绪增强了面对面贸易活动的相关性和影响力。PULL Watersports 的 Jeremy Serwitz 指出："我们是一家'触碰它，感受它'的零售商——没有什么能取代这一点，也没有什么能取代我们与供应商面对面的关系。"

在 2022 年 3 月 Informa Markets 举行的棕榈滩国际游艇展上，与会的客户也表达了同样的观点。"2020 年已经带来了回力棒效应，推动了面对面交易平台的成功。这是我第 22 年作为一名观众或展商来参加这个展览，尤其是今年，这无疑是我经历过的'展会上销售'最成功的一届游艇展。现在，展会结束近一周后，我们的预计销售量增加了一倍多。"Solace Boats 总裁 Todd Albrecht 说。

值得注意的是，80% 的展会客户都是受疫情打击最严重的中小企业，有近 425 万家小企业，而它们定期借助面对面交易平台与符合条件的买家见面。

Fossil Rim 的 Donna Steakley 解释了 Clarion 在美国举办的拉斯维加斯国际旅游纪念品及礼品展览会是今年的第一场展会，以及参加该展会对于下一年春季订单的重要性，"因为我们需要会见供应商，以便亲眼看到针对我们特有地域的艺术品的实际产品和加工。我们看到了客户和销售额的巨大增长，我们知道面对面的环境对于能够签下大量订单是多么重要。"

由全球活动组织者励展博览集团主办的澳大利亚礼品展是澳大利亚礼品和家居用品行业买卖双方的首选交易平台。自 2020 年 2 月以来，该礼品展的悉尼展于 4 月 17 日至 22 日在达令港举行，成为新南威尔士州的首场重大零售活动。该展会吸引了 8 400 多名观众，此次活动的回归受到了来自中小企业主的热烈迎接。Glass on the Grass 的创始人 Lee Drury 说道："我喜欢买卖双方在同一个空间里交流，这是富有魔力的。作为一个开发自有产品的小企业主，能够面对面地会见客户至关重要。当你开发新产品时，能获得客户反馈是非常棒的，而励展礼品展就是最佳场合。多亏励展礼品展，我们得以在澳大利亚各地销售我们的所有产品。2020 年对每个人而言都是艰难的一年，但我们完成了出色的业绩，我们期待着 2021 年的业务也同样强劲。"

全球花卉资源公司的所有者 Joe Clawson 利用由 Diversified Communications 组织的国际花卉博览会来拓展他的供应链。"从一个展位到另一个展位参观，与你前往和出入供应商工厂所花费的时间区别很大……在展会上如果你不喜欢看到的东西，你就继续前进。"

确保健康与安全仍然是展览行业的首要任务，展会行业采取了审慎和协作的方法，以在恢复面对面展会体验时保持一致和严谨。全球领先企业共同制定了一套全行业、经医学审查的安全展会方法，即全面安全标准，并已在 2021 年的全球展会上有效实施，进一步增强人们对经济重新开放后成功、安全地回归 B2B 商业机会的信心。

欧洲是继北美和中国大陆之后的第三大展览市场，也是自新冠疫情开始以来唯一一个贸易展未能恢复规模化的市场。然而，一些展会还是能够推陈出新，例如一对一零售电子商务，这是一个由 Comexposium 组织的活动。"今年正处于危机与复苏之间的边缘。我们的行业已经证明自己能够适应新的形势，现在必须怀着雄心和责任感重新振作起来。'一对一'活动所表现出的领导力和卓越是市场向前发展的最好盟友。"Yves Rocher 集团董事 Bris Rocher 解释道。

全球展览业协会指出，"商业活动行业实乃一个'元行业'：它通过其在世界各地运营的每一个贸易市场和会议场所，在全球范围内赋能每一个行业，从澳大利亚的机床到欧洲的消费品，从亚洲的建筑材料到北美的消费类电子产品。"

除了创造交易市场，以及每年为全球数千个行业和数百万小企业带来数千亿美元的商业销售额外，展览还为区域经济带来巨大的经济价值，每年为主办城市带来 3.03 亿访客，为当地经济注入了可观的收入，包括餐馆、交通服务、精品住宿等相关行业的小企业均受益匪浅。仅拉斯维加斯一个城市就将 114 亿美元的经济影响归功于会展业。

当世界开始从一场对经济和人类都有持久影响的流行病的巨大影响中复苏时，展览成了曙光——在经济和行业社群最需要的时候，提供经济复苏和弘扬重建连接精神的平台。

第六个一年一度的全球展览日和年度庆祝于 2021 年 6 月 2 日举行，它确认了贸易展览在推动全球经济方面所发挥的重要作用。欲了解更多有关全球展览日的信息，请访问 www.ufi.org。

（本文根据互联网新闻报道整理）

思考并回答以下问题：

1. 以上新闻报道中提到了多少场展览会，这些展览会的名称是什么？请查询相关资料，以更全方位地了解这些展览会，如：主办方、展览面积、展馆、参展商、展品及观众来源等。

2. 国际展会的作用有哪些？在新闻中，人们是如何描述展览会的作用的？

3. UFI 是什么组织？在会展行业中发挥着怎样的作用？

4. 采访你身边的人，谈谈他们对线下展览会的认知。

5. 展览会对经济的拉动作用体现在哪些方面？比如说，励展博览集团主办的澳大利亚礼品展，会带动悉尼哪些行业的发展？

6. 国际展览日是哪一天？

Reading 1

Trade Fairs as Part of the Marketing Mix

Trade fairs are a platform for cooperation initiatives; they are meeting places and distribution centers for exchanging products, ideas and know-how. A whole range of marketing aims can be realized at a trade fair. In just a few days the changes of market success of different products and services can be put to the test. Market procedures as changes in direction and speed of future developments will become apparent.

1. General marketing aspects

➢ The marketing function of the trade fair

The basic decision about whether to participate in a trade fair can only be taken after all the questions relating to marketing have been answered. In order to clarify the point "Trade fairs as part of the marketing mix", we shall first have to define the term "marketing".

Marketing can be understood as the planning, coordination and monitoring of all company activities directed towards present and potential markets. These company activities serve the purpose of long-term fulfillment of the customer's needs on the one hand, and the fulfillment of the company's objectives on the other.

To achieve this, the company must bring its whole range of marketing policies into play. The marketing mix consists of product design, adapting to price and conditions and the measuresnecessary for distribution and communication. These tools enable the company to exert an active influence on the sales market.

The trade fair can no longer simply be regarded as an efficient means of distribution; on the contrary, it affects all elements of the marketing mix especially in the case of capital equipment. For the trade fair has changed from being simply a place to buy, it is now increasingly a source of information and communication.

There is great potential for effective marketing in almost every aspect of the mix. When exhibitors take part in a trade fair, they can bring into play their company policies on communication, price and conditions, distribution and products.

Most exhibitors regard participation in a trade fair as an integral component of their marketing mix. Trade fairs serve to fulfill the most diverse company aims.

Marketing at trade fairs means rationalization, because trade fairs can be used for a variety of different functions. Hardly any other marketing tool is capable of combining the detailed presentation of the company and its products with personal customer contact.

Trade fairs are also the source of a multitude of sales leads which are essential component parts of any company's sales policy. This is where market procedures, type and scope of changes as well as direction and speed of future developments really come to light.

Trade fairs are a unique medium with unique possibilities. Unlike an advertisement in a newspaper, a promotional letter, brochure or catalogue which all conveys a purely abstract impression, at a trade fair the product itself is the center of attention. Machines and systems are shown in operation; personal information is directly and inextricably linked to expert technical presentation.

Nor can trade fairs be replaced by highly sophisticated information technology: on the contrary, many products and services have an increasing need to be explained; diverse application possibilities make the ultimate decision more difficult for the buyer. The range of products available is growing all the time. Exchanging experiences and verbal communication are becoming increasingly important. The basis of personal trust between the business partners is therefore one of the most significant factors in making the final decision. In national and in international competition, close customer relationships have taken on a key significance as a strategy for success.

Many marketing objectives can be realized by visiting potential customers at their place of work. Experts have, however, come to the conclusion that, in spite of the expenditure involved in participating in a trade fair, there is on other situation where it is possible to reach so many competent specialists in such a short time as at a trade fair.

The acceptance of a new product or of a prototype can be tested very quickly at a trade fair. The reactions of visitors provide invaluable information for market research.

An additional advantage of participation in a trade fair is the opportunity to maintain contact with regular customers. Instead of the high costs of a visit and the valuable time involved, a short conversation at the trade fair stand will renew the contact and ensure a more intensive business relationship.

Participation in a trade fair must often be seen in conjunction with other marketing tools. For example, if the main aim of participation is to develop existing contacts with regular customers, there must be an intensive campaign to attract them to the trade fair. If, however, the priority is to attract new customers, the advertising campaign must have a broader appeal.

A simple analysis of what happens at a trade fair shows that this marketing tool can achieve a great variety of far-reaching objectives. All the factors mentioned emphasize the importance of trade fairs as part of the marketing mix.

➢ Participation in trade fairs as a company procedure

Just like advertising, sales promotion and public relations, the subject of participation in a trade fair is often hotly discussed throughout all management levels of the company. Insufficient knowledge of the relevant factors, e. g. the selection process and the effect of participation in a trade fair can lead to skeptical reactions or even to rejection of participation in a trade fair. The uncertainties can only be eradicated if participation in a trade fair is regarded as a company procedure, or if it is linked to the dynamic process of a company. Collection and analysis of all relevant internal data (product, product range) and external data (customers, competition) is the first item on the agenda.

After this, the company's own marketing concept must be evaluated with the aid of the data and pre-conditions. The evaluation clarifies whether a trade fair should be used as an additional marketing medium. The evaluation is also required to establish appropriate strategic measures, e. g. aims at the trade fair (see Ch.3, Aims of participation in a trade fair) as well as tactical measures, e. g. employing an individual marketing mix. The realization, that is, the success of the individual aspects, requires appropriate organization, management and monitoring.

2. Trade fairs as part of the communications mix

The process of communication represents an exchange of news and information.

This is also the case at a trade fair, although the exhibitor is at first more in the role of information provider with his stand, products and staff; the visitor initially takes on the role of the information receiver, but later he also become an active participant in the exchange of information. The actual trade fair takes on the role of the medium: thus, communication is one of the central functions of trade fairs and exhibitions.

The instruments of a company's communications policy are advertising, sales promotion, personal sales discussions and public relation. However, market research or the visual image expressed in a company's corporate design can also be included.

The prominent position of the trade fair in comparison to other means of communication is very obvious as a means of communication.

It is clear that the scope of participation in a trade fair, intensive contact between exhibitors and visitors can be achieved a good relationship with customers. Personal conversations between exhibitors and visitors have great value, because this is the only way to develop lasting business relationships and improve existing ones.

It is also the case that a trade fair can convey much more vivid and active information about a product or service than any other component of the marketing mix. The product can usually be seen as well as described. This is particularly significant at capital equipment trade fairs.

On the other hand, trade fairs are in terms of their value as a promotional spectacle and in terms of their availability to the exhibitor. Participation in a trade fire offers a high degree of value as a promotional spectacle. The exhibitor has numerous opportunities to offer the trade fair visitors an impressive experience, e. g. by putting on a "product related" show.

However, the possibilities for exploiting opportunities offered by trade fairs, that is the degree of availability to the exhibitor, are comparatively low, since trade fairs only take place relatively infrequently on a rotational basis, and deadlines for registration must be observed. This means that the exhibitor must allow for a longer-term planning period if a trade fire is to be used.

Great importance is attached to a trade fair in comparison, to other media because of its multifunctional character. No other medium can be employed in such an individual way, and no other situation offers the opportunity of such direct communication with customers, for the purpose of creating a need for information, or satisfying the already existing need for information. Specific advantages of your product and/or company, such as reliability, good after sales service and high product quality, can be presented quite clearly. During participation in the trade fair, important information about, sales promotion or advertising, for example, goes back to the company where it should be used to good advantage.

Participation in a trade fair helps a company to reach more potential customers and to create a more favorable impression on existing customers. It is also possible to become aware of changes in the customer profile and in buying behavior more quickly and more directly within the scope of participation in a trade fair.

3. Trade fairs as part of the price and conditions mix

Important aspects of the price and conditions mix include price, credit, discount, payment and service.

The relevant spheres of influence for the development of an individual price and conditions mix are, in particular, exact knowledge of the customer profile, of the size of companies, of the locations and of the delivery distances involved. The company can find out the necessary information in a conversation with the customer, for example.

Participation in a trade fair contributes towards a new conception of the existing price and conditions mix, and if desired, new areas of the market can be sounded out. The following points should be taken into consideration: packaging, freight and insurance costs; costs for after sales service and customer service; existing price calculations; conditions of payment; payment, quantity and special reductions; conditions of delivery; conditions of cancellation; questions relating to customers' prompt setting of accounts and credit-worthiness should also be considered. The price and conditions mix must be arranged so that company aims can be achieved and company profits assured.

4. Trade fairs as part of the distribution mix

The distribution mix can be represented as follows: sales organization, distribution channels, storage and transport. An explanation of the aspects of the distribution mix shows to what extent participation in a trade fair can be regarded positively.

There must be an investigation into whether further reorganization or other changes to the existing sales organization are necessary, for example, restructuring the sales force, recruitment of dealer and sales representatives, looking for cooperation partners for storage and transport.

Another consideration is whether the existing distribution channels need to be changed qualitatively or quantitatively.

5. Trade fairs as part of the product mix

Important aspects of the product mix are as follows: product quality, product range, brands and product design.

One important consideration is to what extent the product range of the company should be on show at a trade fair. In order to review the product mix, it is necessary to be aware of the current market cycle of the product or specific market that is to be displayed. The range can be extended if desired in order to be able to offer an even better presentation at a trade fair where a new product — a trade fair launch — is to be shown. The product design should be up-to-date and thus easily marketable; the same applies to the packaging.

Should the company's product be branded merchandise or should there be trademarks used, thus must be explained.

Individual elements of the product mix, and thus the product itself, can be tested by participating in a trade fair. By talking to a user, the acceptance of the product can be tested and thus the company can gain valuable stimuli for its product and product range policy.

6. A word about competitors

The analysis of the strengths and weakness of the competition provides more information in respect of the decision about possible participation at a trade fair. The aims of this analysis of the competition might be: more informed assessment of your own market position; to learn from the market behavior of competitors; better assessment of your own development.

It is important to establish who exactly should be regarded as the competition as well as direct competitors who manufacture the same, or similar, products, it is also necessary to include companies that use the same production processes, or offer substitutes for your company's products.

It is possible to distinguish between individual competitive markets on the basis of, for example, products and services, and according to geography.

Competition circumstances for each individual market can then be determined with the aid of various criteria, e. g.: company image, company location and factories, rang of services, manufacturing capacity, research and development activities, marketing strategy, marketing mix, advertising budget, distribution network, profitability, development trends.

Due to the meeting of different companies with a similarly structured range of products, participation in a trade fair will enable you to see the market with additional clarity.

Within the scope of medium and long term company planning, participation in a trade fair can serve to clarify the future position of a company in the market.

The cost to profit ratio can also be improved by participating in a trade fair.

In the case of participation in an international trade fair, however, three criteria should be fulfilled. Participation usually makes sense where: the company's sales are not limited to one region; sales are based on a sufficiently broad customer base; the product, or the service, shows a high degree of know-how. Mass-produced goods or everyday products are hardly likely to meet with a positive response at a trade fair. Visitors come in the expectation of discovering new products and seeing technologically highly advanced products, or special products, "in the flesh".

Individual gaps when answering the catalogue of questions may lead to uncertainty about participation in a trade fair, but they should not have a detrimental effect on the overall decision. The answer to most questions which derive from the marketing mix will give a basic structure for an individual marketing mix and thus provide the answer to the question about whether to participate or not.

The effects of participation in a trade fair within the scope of the company's overall plan for marketing policy must also be taken into consideration. Participation in a trade fair results in positive cooperative effects for a company. Participation in a trade fair can above all lead to success when there is a conscious effort to coordinate it with the other elements of the marketing mix.

Participation should be agreed for a period of time spanning at least three events. Taking part in one event only gives a distorted picture, and most importantly of all, the opportunity to intensify those first contacts made at the central meeting place and competition arena of your branch of industry, the trade fair, would be missed.

(This article is edited according to Internet information)

NEW WORDS AND PHRASES

platform   n. 平台

cooperation   n. 合作, 协作

initiative   n. 倡议; 新方案

distribution   n. 分配

know-how   n. 诀窍; 实际知识; 专门技能

procedure   n. 程序, 手续; 步骤

apparent   adj. 显然的; 表面上的

bring into play   开始起作用

integral   adj. 完整的, 整体的

component   n. 成分; 组件

rationalization   n. 合理化

combine   v. 使联合, 使结合

come to light   为人所知

insufficient   adj. 不充足的

skeptical   adj. 怀疑的

rejection   n. 拒绝

uncertainty   n. 不确定, 不可靠

eradicate   vt. 根除, 根绝

agenda   n. 议程; 日常工作事项

concept   n. 观念, 概念

evaluate   vt. 评价; 估价

appropriate   adj. 适当的

tactical   adj. 战术的; 策略的

stand   n. 展位, 展销台

staff   n. 职员; 员工

initially   adv. 最初地

promotional   adj. 促销的

availability   n. 可用性; 有效性; 实用性

comparatively   adv. 比较地; 相当地

infrequently   adv. 很少发生地; 稀少地

deadline   n. 截止期限, 最后期限

registration   n. 登记; 注册

observe   v. 遵守

multifunctional   adj. 多功能的

reliability   n. 可靠性

after sale service   售后服务

aware   adj. 意识到; 有……方面知识的

sphere   n. 范围

take into consideration　考虑到

package　n. 包装

freight　n. 货运;运费

insurance　n. 保险;保险费

calculation　n. 计算

cancellation　n. 取消;删除

prompt　adj. 敏捷的,迅速的;立刻的

positively　adv. 肯定地,明确地

recruitment　n. 补充

launch　vt. 发行　n. (产品的)上市

marketable　adj. 有销路的,可销售的

profitability　n. 收益性;收益率

profit　n. 利润;利益

ratio　n. 比率,比例

criteria　n. 标准,条件(criterion 的复数)

catalogue　n. 目录

detrimental　adj. 不利的;有害的

derive　v. 源于;得自

conscious　adj. 有意的,刻意的

span　v. 持续,贯穿

distorted　adj. 歪曲的;受到曲解的

 NOTES

1. Trade fair，贸易展览会。展览会有多种英文表达方式，如：exhibition，fair，show，exposition，国际上展会的名字通常会使用简写或者全称，如，德国法兰克福春季消费品展览会（Ambiente）、德国法兰克福家用纺织品展览会（Heimtextil）。

2. Marketing mix，市场营销组合。其是企业市场营销战略的一个重要组成部分，是指企业针对目标市场的需要，综合考虑环境、能力、竞争状况，对自己可控制的各种营销因素（产品、价格、分销、促销等）进行优化组合和综合运用，使之协调配合，扬长避短，发挥优势，以取得更好的经济效益和社会效益的整体性活动。

1964 年，麦卡锡（McCarthy）提出了 4Ps 营销组合，即产品（Product）、价格（Price）、渠道（Place）和促销（Promotion）。1981 年，布姆斯和比特纳（Booms and Bitner）在此基础上提出了 7Ps 营销组合，即在 4Ps 的基础上增加了人（People）、有形展示（Physical Evidence）和过程（Process）这三项元素。7Ps 也构成了服务营销的基本框架。

3. Capital equipment，资本设备。其是指企业用于提高生产率或者进行生产现代化改造的设备。

4. Communications mix，营销中的沟通组合。其包括公司与客户沟通的各种方式，如广告、销售促进、公共关系与宣传、人员推销、直接营销和互动营销等。

5. Price and conditions mix，价格和条件组合。其包括价格、信用、折扣、付款和服务。

6. Distribution mix，分销组合。可以表示为销售机构、分销渠道、储存和运输。

7. Product mix，产品组合。其包括产品质量、产品范围、品牌和产品设计。

EXERCISES

Ⅰ. Answer the following questions.

1. What are the functions of a trade fair?

2. Interpret the marketing function of a trade fair.

3. Interpret the communication function of a trade fair.

4. Interpret trade fairs as part of the price and conditions mix.

5. How can a company set up a distribution channel through taking part in a trade fair?

6. What is a competitor?

Ⅱ. Please translate the following English sentences into Chinese.

1. Marketing can be understood as the planning, coordination and monitoring of all company activities directed towards present and potential markets.

2. Trade fairs are also the source of a multitude of sales leads which are essential component parts of any company's sales policy.

3. An additional advantage of participation in a trade fair is the opportunity to maintain contact with regular customers.

4. However, market research or the visual image expressed in a company's corporate design can also be included.

5. Participation in a trade fair can above all lead to success when there is a conscious effort to coordinate it with the other elements of the marketing mix.

6. Participation usually makes sense where: the company's sales are not limited to one region; sales are based on a sufficiently broad customer base; the product, or the service, shows a high degree of know-how.

7. The aims of this analysis of the competition might be a more informed assessment of your own market position.

8. The aims to be pursued at the trade fair are consistently derived from the individual marketing aims.

9. Any resulting order is termed an indirect trade fair purchase order.

10. This means that the exhibitor will either select a suitable trade fair according to his established aims, or will vary the aims according to the trade fairs available.

Ⅲ. There are 10 sentences in this section. Beneath each sentence there are four words or phrases marked A, B, C and D. Choose one word or phrase that best completes the sentence.

1. Which of the following is the best translation of the phrase "marketing mix"? (   )

A. 市场混合   B. 营销混合   C. 营销组合   D. 销售组合

2. The phrase "sales leads" most probably means (   ).

A. sales chances   B. sales leaders   C. sales personnel   D. sales of leads

3. The phrase "customer profile" can be translated into (   ).

A. 客户形象   B. 客户印象   C. 客户外形   D. 客户概况

4. The communication via language is called (   ).

A. verbal communication   B. body language

C. tele communication   D. mobile communication

5. The phrase "old customers" is also called (   ).

A. regular customers   B. traditional customers

C. frequent customers   D. elderly customers

6. The phrase "come to light" probably means (   ).

A. become not so heavy   B. not difficult

C. not in the dark   D. come to be known

7. Which of the following is the closest in meaning with the phrase "marketing objectives"? (   )

A. aim of marketing   B. objective methods of marketing

C. marketing the goods   D. marketing objects

8. Which of the following is the closest in meaning with the phrase "potential customers"? (   )

A. possible clients   B. Hidden customers

C. foreign customers   D. potent customers

9. What is best Chinese translation of the phrase "know-how"? (   )

A. 知道怎样   B. 诀窍   C. 方法   D. 途径

10. What of the following is the closest in meaning with the phrase "in conjunction with"? (   )

A. together with   B. connected with

C. conjuncture of   D. on condition that

Ⅳ. Writing

Please write a reply to QQQuan Technology Co. , Ltd. with the following particulars:

1. Acknowledge the receipt of their letter of March 8.

2. Agree to their proposal of establishing trade relations with you.

3. Commodity inspection will be handled by the bureau concerned in New York.

4. Glad to meet the General manager in National Hardware Show in Las Vegas in May.

Reading 2

Aims of Participation in Trade Fairs

Before making the final decision about participation in a trade fair, an analysis of your company's situation and a clear definition of your own starting point are indispensable. Experts emphasize again and again the importance of establishing the communication, price and conditions, distribution and production aims before participation.

When doing this, the company's aims as established within the scope of medium-term company planning can be seen as a starting point for a plan for committed participation in a trade fair as part of the marketing mix. The aims to be pursued at the trade fair are consistently derived from the individual marketing aims.

In the context of capital equipment, it is often argued that trade fairs have little or no sales value—that at best they can be useful for sales preparation. This cannot lead to the conclusion that communication should be regarded as the only aim, since the difference between an initial or contract visit to a customer by a salesman—which does not usually lead to an immediate order either—and a trade fair, is considerable.

Corresponding to the multifunctional nature of trade fairs, a whole package of marketing aims can be realized. For communication, the following applies: even if only the sales process is in the foreground, the company can still do useful public relations work at the same time. It is also possible to observe the competition for the purpose of market research. The company introducing a new product can also take advantage of the trade fair for general sales promotion and advertising. The breadth of possibilities available is a direct result of the opportunity which the trade fair and no other medium-offers for personal contact with a large number of people interested in your branch of industry.

Establishing the most important trade fair aims influences the whole organizational preparation right through to the completion and monitoring of participation aims. The following grouping of participation aims needs to be considered with a flexible attitude. This is an example of how communication aims can also serve the product mix.

Primary participation aims are: to encounter new markets (discover niches in the market place); to examine your competitiveness; to assess export chances; to inform yourself about the situation of the branch of industry; to exchange experiences; to initiate cooperation arrangements; to participate in specialist events; to recognize development trends; to interest new markets in your company or product; to combine participation in a trade fair with complementary measures (special events, seminars, tours of the factory); to meet the competition (which competitor exhibits at which trade fair?); to increase profits.

Communication aims are: to develop personal contact; to meet new groups of customers; to increase company prominence; to increase the effectiveness of company advertising among customers and the public at large; to complete the index of customers; to consolidate press relations; to discuss requests and requirements with customers; to collect new market information; to put the corporate design plan into action; research and sales training through an exchange of experiences.

Price and conditions aims are: to present a convincing range of services to the market; to sound out the room for manoeuvre as regards pricing.

Distribution aims are: to expand the distribution network; to estimate the effect of elimination of the trading level; to look for new agents.

Product aims are: to test the acceptance of the product range in the market; to launch prototypes; to assess the success of a product launch on to the market; to present product innovations; to expand the product range.

Let's come to talk about the aims of the exhibitor regarding visitors. The primary aims of a specialist visitor provide the exhibitor participating at a trade fair with guidelines for strategic planning. This is extended into the area of tactics; it is then possible to talk about the exhibitor's aims being determined by visitor−orientated criteria.

Here is a selection of visitors' aims: to gain a general view of the market, including related specialist areas; to assess the situation and perspectives of the market; to compare prices and conditions; to look for specific products; to see new products and possible applications; to recognize market trends; to become informed about the technical function and nature of certain products or systems; to find information on the solutions to current problems; to visit conferences and special shows; to learn; to gather suggestions regarding his company's own product and range design; to develop, or make, business contacts; to place orders and negotiate contracts; to seek out contact in similar companies; to assess the benefits of possible participation as exhibitor.

The weighting of the individual participation aims depends decisively on the desired and the possible trade fair. Two examples can serve to clarity this:

At trade fairs for consumer good, e. g. toy, fashion, leather good, the main attraction is the placing of orders. The visitors to these events are mainly customers who want to order goods or designs for the next season. Characteristic of these fairs is that the products are bought according to a fixing rotation and that the buyer can usually make an immediate decision without having to consult the management of the company involved.

Immediate profit is comparatively unimportant at international trade fairs for capital equipment. This equipment is relatively complex and the order is often only placed after a long period of negotiations, since considerable technical problems must be solved. The conditions are also the center of intensive negotiations. Several people authorized to make decisions are involved with the purchase.

The negotiations take place in the period after the trade fair. Any resulting order is termed an indirect trade fair purchase order. Purchase orders for capital equipment are frequently prepared in advance so that the contract can be signed at the fair.

As far as the weighting of the participation aims is concerned, this means that the exhibitor will either select a suitable trade fair according to his established aims, or will vary the aims according to the trade fairs available.

In the case of events which are based more on information and consultancy, it is more difficult to define the aims in terms of the size of profit made.

To conclude, one more important point must be considered: what is the level of financial means available for participation in a trade fair and for the absolutely essential additional measures (e. g. pre-trade fair advertising)？ Participation in a trade fair is often ruled out because of the costs involved, without the multifunctional nature of trade fair having been considered during the decision-making process. The question is, therefore, whether your company is prepared to undertake financial restructuring in order to accommodate a trade fair regarded as necessary, even though it already has long-established aims.

(This article is edited according to Internet information)

 NEW WORDS AND PHRASES

indispensable　adj. 不可缺少的；绝对必要的

committed　adj. 坚定的；承担义务的

pursue　v. 追求，实现

consistently　adv. 一贯地；一致地

contract   n. 合同

considerable   adj. 相当大的; 值得考虑的

in the foreground   在前景中

public relation   公共关系

breadth   n. 宽度, 幅度

flexible   adj. 灵活的

encounter   vt. 遇到

competitiveness   n. 竞争力

complementary   adj. 补足的, 补充的

effectiveness   n. 效力

index   n. 指标; 指数; 索引

consolidate   v. 巩固

convincing   adj. 令人信服的; 有说服力的

manoeuvre   n. 策略

expand   v. 扩大, 增加

elimination   n. 消除; 除去

prototype   n. 原型; 样机

innovation   n. 创新, 革新

guideline   n. 指导方针

visitor-orientated   adj. 以顾客为导向的

perspective   n. 远景

negotiate   v. 谈判, 商议

assess   vt. 评定; 估价

decisively   adv. 决定性地

rotation   n. 轮流

consult   v. 咨询

consultation   n. 咨询

consultancy   n. 专家咨询

authorized   adj. 经授权的

financial   adj. 金融的; 财政, 财务的

absolutely   adv. 绝对地; 完全地

essential   adj. 基本的; 必要的

accommodate   v. 使适应; 供应

Reading 3

Manning the Stand

The trade fair stand creates a positive environment for product and personal information by means of its architecture and design. Competent stand personnel and efficient, functional running of the trade fair ensure success.

Stand personnel: the better motivated and qualified the stand personnel, the greater the chances of good sales results and new contacts. Purposeful selection and intensive training of the stand personnel are just as important as an effective presentation of the products.

Personnel planning and selection: the selection of suitable employees for work at a trade fair is based both on their specialist knowledge and their personal qualities.

Qualities of the Stand Personnel: outstanding theoretical and practical specialist knowledge; ability to deal with people and openness; confident an proficient manner; articulate expression; flexibility; knowledge of foreign languages; experience at trade fairs; stamina (health); willingness to travel.

Temporary staffs are available on site for a whole range of activities on the stand. These include stand assembly and disassembly catering and entertainment, as well as interpreters may usually be arranged on an hourly basis by the organizer.

According to the size of the company, the stand personnel should include: company representative (member of the board of directors, managing director); stand management (responsible for running the stand); technical staff (consultation demonstrations); sales staff (sales, conditions of delivery); staff responsible for trading countries abroad (export discussions); interpreters; press agent; information personnel (stand information); service personnel (office, catering, waiting staff, security, cleaning).

Motivation and training: at a trade fair the company as a whole is under scrutiny. Every member of the trade fair team must be willing to give his or her best before or during the trade fair as well as in the follow-up stage. Employment at a trade fair is not a reward, it is hard work, for which the staff must have intensive preparation.

The more comprehensively the stand personnel are informed about the aims of participation and the more clearly all individual's duties are defined, the better each employee is able to fulfill the requirements. Stand personnel who feel properly prepared and informed make a considerable contribution to the smooth and successful running of the stand. They should be informed about the following: the company's own range of products and services; price and conditions; the competition and competitor's range of supply; the target group; the visitor profile of the trade fair; important customers and interested parties; how to record each conversation with a visitor; the layout of the stand and the duty roster; the importance of the trade fair for the branch of industry; the location of the trade fair and the trade fair grounds.

It is frequently the case that employees only have limited practical experience in dealing with trade fair visitors. The trade fair team must therefore be prepared and trained for this task, especially in how to conduct discussions, present arguments, and answer questions. There is a variety of special seminars, publications and videos on the subject of trade fair training.

If as many contacts as possible are to be made, the most successful employees are those who attract the attention of the visitor. Every visitor is a potential customer. The art is to win him over. A prerequisite for this art is the ability to make an active approach to the visitor. Experts estimate, however, that 50% to even 90% of all conversations are initiated by using the conversation killer "Can I help you?" This phrase frequently kills the conversation stone dead in a matter of seconds.

The customer also wants to be at center attention. This can be achieved verbally if the member of staff speaks from the point of view of the customer, i. e. , "you receive" instead of "we supply", "Here you can see" instead of "I will now show you".

The most shocking results were obtained from observation made at capital equipment trade fair in particular. Up to 70% of visitors are not approach at all. 80% of sales staff ends the conversation if the visitor has a cold manner.

Training for trade fair has the following points of emphasis. How is the interest of the visitor aroused? How and when are they approached? How should their name and address be requested and written down (visitor records)? How should member of staff behave towards the general public?

The training has the following objective: it is important to communicate to the visitors that the staff is approachable at all times. It is especially important to avoid behavior which will discourage visitors from coming to look at the stand (reading a newspaper, involved conversations with friends or colleagues). Trade fairs are live events; there are no second chances.

It is often the case that visitors will only enter a particular hall once during their visit and are interested in one particular stand for a short while. If the stand personnel don't signal willingness to communicate, a potential contact will be lost.

It is also very important to be aware of dress, appearance and posture. A uniform for the stand personnel makes it easier for visitors to find assistance. It is often enough to have certain clothing accessories (tie, scarves) to make the stand personnel more easily recognizable. It goes without saying that name badges that are big enough to be read easily should be worn.

Conducting conversations: when visitors enter the stand, they must be allowed sufficient time to look around. The stand employee should notice what it is that interests them. At the same time, they can wait for a suitable moment for the initial contact. When the visitor is being welcomed, members of staff should introduce themselves and offer comments on the relevant exhibit. Visitors known to the staff should be approached immediately and be greeted by name.

During a conversation, excessive insistence is to be avoided at all costs. The ability to listen is what is required here. Motives, criticisms, intended applications, quality requirements and how quickly the decision to purchase must be made should all be found out by the use of relevant questions. By enquiring about the level of professional competence and the decision-making power the visitor has within the company it is possible to find a basis for mutual understanding. If possible, the best reaction to objections and superficial arguments is to be sensitive and reply with concrete solutions.

At the end of the conversation, further contact should be arranged, if possible, e. g. an appointment for a visit, or sending a concrete quotation or technical details.

When the time comes to fill in the visitor record, all requests should be noted at once. Otherwise, it is easy to forget something in the hectic atmosphere of a trade fair. Exact information for the follow-up work avoids the situation of the interested party being given contradictory information at the next meeting.

Stand organization: the stand manager is responsible for the smooth running of the trade fair-stand both externally for visitors and internally for staff.

The stand manager must have a number of qualities and flair for dealing with people of greatly differing temperaments. The stand manager must: have experience of trade fairs and exhibitions; be able to make, and enjoy making, decisions; have motivating and leadership qualities; have a talent for organization and improvisation; have a sense of responsibility; have a smart appearance; be confident; be articulate; be willing to conduct discussions and negotiations; have basic technical and commercial knowledge; have a good memory for people.

A deputy to the stand manager should be appointed in good time (in case this is required at short notice or in the case of absence due to unforeseen circumstances).

Organization of the trade fair stand: the stand manager should be satisfied themselves well before the start of the trade fair that the whole stand will be built according to plan, that the furnishings, design and product captions are in order, that publicity material and catering are available and that all connections and equipment are in working order.

On the evening before the trade fair, the stand personnel should receive their instructions and the procedures at the stand should be explained: introduction of the members of the team, including any outside personnel, information about the trade fair, information about the aims at the trade fair, special events and occasions during the trade fair, explanation of important documents, explanation of the visitor record forms, instructions on how to entertain customers, instructions for press agents, planning of the duty roster, code of behavior.

The duty roster establishes who is responsible for individual tasks, e. g. for making sure the brochure racks are full, for keeping the discussion booths or groups of chairs clean during the day, for the catering. It also determines a rota for breaks during the day.

The daily discussion of the situation in the morning or in the evening serves to inform all members of staff about successes attained at the trade fair and about any special details for the following day (important visitors or events). It is also possible to discuss weaknesses in the procedure and find short term solutions. A similar process of criticizing the adopted approach is also to be recommended for the end of the trade fair.

A well-organized and well-managed trade fair team ensures that the trade fair stand is clean and tidy at all times; that no bottle-necks occur with consumer goods and provisions(publicity material, catering); that all technical facilities at the trade fair stand are fully functional; that stand procedures and working hours are kept to; that the atmosphere at the trade fair stand is always friendly and relaxed; that the stand manager always knows exactly where employees are at any one time; that conversations with visitors are written down and analyzed.

Catering can be organized even on a small stand. Soft and alcoholic drinks may be served. If cakes, biscuits or nibbles are served, they should be fresh and presented in an attractive way. It is also necessary to provide sufficient crockery and cutlery, and washing-up facilities.

If the catering is associated with the region of origin of the company e. g. Franconian wine, Munich sausage, or Westphalian ham, this acts as a memory aid.

At all trade fair grounds there are outside contractors who deliver drinks or food to the stands.

Visitors' records: in order to be able to carry out effective follow up work and make concrete comments on the success of the trade fair, visitor records are essential. Pre-printed forms can help reduce the work involved and therefore the time required of the employee to fill them in. Enquiries can only be dealt with promptly if these report forms are filled in accurately. Experience shows that after a short running-in period, employees find these forms are a valuable working tool.

It must be established beforehand exactly which sorts of conversations are worth recording as a general rule, the report forms are only filled in if the visitor has a serious interest in the product. Short pieces of detailed information which do not include an address can possibly be included in lists showing interest expressed according to product or subject group. These brief conversations can be also given important information on the acceptance of the exhibits.

Trade fair and market information: the employees working on the stand can do some market research as well as looking after the stand. Information about the products, stand design and activities of competitors are useful starting points. Walking round the trade fair serves to motivate member of staff and give them some extra training. The publications available at the trade fair should also be evaluated: trade fair catalogue, brochures about special events, lecture summaries, special editions of specialist magazines, competitors' brochures and publicity material, organizer's questionnaires.

Cleaning and security: it goes without saying that the stand should be kept clean at all times. The daily cleaning can be done by members of staff or one of the organizer's outside contractors can be employed to do it. In addition to this, the employee responsible should ensure that everything is clean and tidy during the day. Overflowing ashtrays, brochures lying around and stale biscuits quickly lead to people drawing the wrong conclusions about the exhibitor's overall service.

Security at the trade fair stand must be properly organized even during assembly and disassembly. Experience dictates that these times are chaotic and hectic; valuable exhibits should not therefore be left unattended.

This also applies to the daily running of the stand. In particular in the case of trade fair with large amounts of visitors, adequate security arrangements should be made for valuable exhibits.

Trade fair halls usually have their own security arrangement at night. It is also possible to employ your own company security for stands with very valuable exhibits.

Concluding business: directly after the trade fair has finished, there should be a concluding-discussion for the benefit of members of staff. While things are fresh in everyone's minds, a whole range of matters can be aired and their significance evaluated for the next trade fair. A written report can then contain recommendations for future presentations.

The disassembly process can begin after the event has officially ended. Visitors should not arrive on the last day to be greeted by the sight of a half-empty stand. Atmosphere of departure does not exactly make an interested visitors fell welcome.

Time can be saved and stress avoided if the disassembly process and the transport have been organized well in advance.

(This article is edited according to Internet information)

NEW WORDS AND PHRASES

manning   n. 人员配备

man   vt. 为……配备人员

architecture   n. 建筑风格；建筑样式

competent   adj. 胜任的；有能力的

purposeful   adj. 有目的的；有决心的

theoretical   adj. 理论的；理论上的

articulate   adj. 发音清晰的；口才好的

stamina   n. 毅力；精力；活力；持久力

catering   n. 提供饮食及服务

the board of director   董事会

managing director   总经理，常务董事

press agent   新闻广告员

scrutiny   n. 详细审查；监视

comprehensively   adv. 包括一切地

duty roster   执勤表；轮值表

seminar   n. 讨论会，研讨班

prerequisite   n. 先决条件

stone dead   完全被摧毁的

verbally   adv. 口头地，非书面地

badge   n. 徽章；证章；标记

decision-making   adj. 决策的   n. 决策

hectic   adj. 兴奋的，狂热的

flair   n. 资质；鉴别力

improvisation   n. 临时准备

deputy   n. 代理人，代表

unforeseen   adj. 未预见到的，无法预料的

circumstance   n. 情况；事件

caption   n. 标题；说明

crockery   n. 陶制餐具

cutlery   n. 餐具；刀具

Franconian wine   法兰克葡萄酒

Munich sausage   慕尼黑香肠

Westphalian ham   威斯特伐利亚熏腿

sort   n. 种类

overflowing   adj. 溢出的；充满的

ashtray   n. 烟灰缸

dictate   v. (经验或常识) 使人相信

chaotic   adj. 混沌的；混乱的，无秩序的

conclude   v. 达成(协议)

business   n. 生意; 交易

 ENGLISH FOR WORKPLACE COMMUNICATION

Sample Dialogue: Establishing trade relation in a trade fair.

Situation: The following conversation is between Mr. Smith (B), a potential importer from the USA, and Ms. Lee (A), director of ABC Garment Co, Ltd. , China. Ms. Lee is accompanying with Mr. Smith to visit the booth in Magic show in Las Vegas, USA.

B: Would it be possible for me to have a closer look at your samples, Ms. Lee?

A: Sure. I'd love to show you, Mr. Smith. This way, please.

B: Thanks.

(3 minutes later)

A: Here we are. This is our booth. Isn't it nice, Mr. Smith?

B: Yeah. You've surely got a large collection of sample garments here.

A: Definitely. In Hebei province in China, we are the biggest exporter with a wide range of garments to the rest of the world. And the global demand of our products is still on steady increase.

B: Just as what I got to know. Frankly speaking, our clients seem to be quite addicted to Chinese natural fiber garments. Actually, in the past 10 years, your textile exports to the USA have doubled in spite of the strict quota! So incredible!

A: This is not an exaggeration at all. The quality of our products is as good as that of our competitors, well, our prices are pretty attractive. So, Mr. Smith, which items are you interested in?

B: Silk garments are of special interest to me, particularly the silk blouses and skirts. As your silk blouses are among the most popular ones in our market, I'm going to place a large order in a couple of days.

A: Sounds great! How about our chemical fiber products?

B: I think they will also find a good marketing in our country. Will you show me some samples?

A: Sure. This way, please! Our chemical fiber products cover a wide range, such as ties, belts and shirts.

B: Shirts are more saleable in our market than the others. Could I have your latest catalogues or something that tells me about your company?

A: Certainly.

B: Well, it's already lunch time. Shall we talk about the details over lunch?

A: OK, I'll treat you with the local cuisine.

B: Thank you very much! You are so kind to me!

A: It's my pleasure.

UNIT 2

PREPARING A TRADE FAIR

本章导读

　　参加国际展览会是公司开拓市场的战略组成部分，选择展览会是筹备工作的第一步。本章重点学习如何选择一场合适的展览会，在选择展览会之前要学会市场细分和寻找目标客户。不同国家和地区的展览会覆盖不同市场，客户需求也各不相同。在筹备过程中，展位设计、参展人员培训等都是重要环节。通过本章的学习，熟悉如何筹备一场展览会。筹备工作涉及很多环节，包括展品、物料、客户邀约、产品报价单、广告宣传等。一场展览会筹备工作的充足与否直接决定展出效果。

知识精讲

展会知识

2019 年汉诺威工业博览会开幕

　　新华社德国汉诺威 3 月 31 日电（记者：沈忠浩、任珂）以"融合的工业——工业智能"为主题的 2019 年汉诺威工业博览会（以下简称汉诺威工博会）于 3 月 31 日晚拉开帷幕，人工智能、第五代移动通信技术（5G）应用等将成为关注重点。

　　3 月 31 日，在德国汉诺威，瑞典首相勒文在 2019 年汉诺威工博会开幕式上致辞。本届汉诺威工博会的主宾国是瑞典。作为全球规模最大的年度工业展会，本届汉诺威工博会重点关注人工智能、5G 与"工业 4.0"的结合，将展出超过 100 个机器学习的应用实例，较 2018 年明显增多。同时，汉诺威工博会今年首次搭建大型 5G 测试平台，多样化的工业应用将在一个真实的 5G 网络环境中得到展示。

　　德国总理默克尔在当晚的开幕式上呼吁德国工业企业勇于创新，通过适应环境和自我改变保持竞争力。

　　汉诺威工博会创立于 1947 年，现每年举办一届，今年共吸引来自全球 75 个国家和地区的 6 500 家参展商，海外展商比例约 60%。其中，中国展商数量仅次于东道主德国。本届工博会将持续至 4 月 5 日，参观人数预计超过 22 万。

　　思考并回答以下问题：

　　1. 查询相关资料，了解汉诺威工业博览会是怎样的一场展会？

2. 河北省邢台市有一家企业生产轴承系列产品，这家企业想开拓欧盟市场，是否可以选择参加汉诺威工博会？说说你的理由。

3. 如果河北企业要参加汉诺威工博会，他们应该提前做哪些准备工作？

4. 这家企业的销售人员认为，在德国参展，观众和买家大多是德国人，你如何用数据消除参展企业的这个误解？

5. 除德国汉诺威工博会外，其他国家还有哪些有影响力的工博会？比如日本、美国、非洲和中东市场等。

📖 Reading 1

Step-by-Step Guide to Successful Trade Show Preparation

1. Why exhibit

Trade shows are a popular marketing component of business. Trade shows allow companies to interact and forge new partnerships with a diverse group of potential buyers and clients, all within one location. Many businesses recognize these advantages and incorporate trade show exhibition as an important part of their company's export strategy.

Fact: 76 percent of companies indicated their return on investment is the same or higher at participation in international trade shows compared to their participation in domestic trade shows. Source: EXHIBITOR Magazine, 2019.

Trade shows are:

A global marketplace for buyers and sellers to meet, do business and make sales;

A forum to increase your profile and differentiate yourself from competitors;

A unique, highly effective, non-mass media marketing tool to promote existing products or to launch new ones;

An opportunity to network and find new clients and generate new leads;

A venue to find new investment opportunities;

A cost-effective way to gather market research and learn about industry trends.

2. Building your marketing strategy

Once you have decided to participate and exhibit at a trade show, it is very important to create a marketing strategy to promote your brand both leading up to and during the event. Having a marketing strategy in place will allow you to take steps to increase your profile at an earlier stage and differentiate yourself from competitors. You want to give attendees a reason to visit your booth and do business with you at the event.

Fact: Attendees are more receptive and engaged with marketing messages at special events, increasing the likelihood of short-term marketing at trade shows successfully converting to sustained brand recognition and value. Source: Center for Exhibition Industry Research.

➣ Identify your objectives and target audience

Common objectives include increasing brand and product awareness, expanding market presence, meeting new clients, researching industry trends, launching or testing a new product, and maintaining relationships with existing clients and partners. Defining your target audience will ensure a more tailored and efficient strategy.

➤ Budget

Marketing is just as important as participating at the event and should be included as part of the overall budget. If exhibiting internationally, be mindful that some marketing costs may be incurred overseas and will be affected by foreign exchange rates.

➤ Select the right marketing tools

Common marketing methods include:

Direct marketing to consumers;

Individual appointments and meetings;

Advertising and listing in event directories;

Media and public relations packages;

Marketing and sponsorship of events and functions at the trade show;

Social media;

Social media tools.

Effective use of social media can boost your trade show strategy by providing real-time client interaction and tailored content for your target audience. It can provide attendees with new channels to learn about, communicate and recommend your company and products.

Fact: 91 *percent of exhibiting companies currently incorporate social media into their face-to-face marketing mix. Source: EXHIBITOR Magazine,* 2018.

Businesses and consumers are increasingly incorporating social media as part of their marketing strategy to communicate, share and promote their brand, services and products. Popular tools include Facebook, Twitter, Instagram, LinkedIn, Flickr, and YouTube.

Leveraging social media:

Decide on the best social media tools to reach your target audience. Find out what social media tools and online forums they are using.

Look into what social media tools trade show organizers are using and connect with them.

Develop engaging and innovative content to differentiate yourself from the crowd.

Ensure that the social media tools for sharing and embedding your content are displayed prominently, strategically located, and easy to access whether it is on promotional material or on your website.

Regularly track and monitor the content viewed and feedback received through social media. Your clients are your extended sales force and acknowledging their feedback can help further develop and improve your business.

Use social media tools to interact with existing and prospective clients. Engage clients to help share your story about who you are as a business and a brand. Remember, social media is supposed to be social.

➤ Implement the strategy

The strategy must then be continually monitored and each marketing activity recorded and evaluated on outcomes achieved and lessons learned. Establishing a timeline for execution is critical to stay on track.

3. Choosing the right show

There are hundreds of trade shows around the world to choose from. Many companies spend time and money on travel and booth creation and fail to spend resources on researching shows, and setting goals. Research and careful assessment will ensure that the trade show selectedaligns best with your objectives and goals.

For Example, use Trade Commissioner and Agriculture and Agri-Food Canada (AAFC) Regional Office expertise. Trade Commissioners are stationed around the world to promote the Canadian agriculture and Agri-Food Sector. Trade Commissioners located in your international market of interest, as well as staff at regionally located AAFC offices throughout Canada, are familiar with conducting business within specific markets, the key stakeholders, as well as international trade events held in specific locations.

➢ Identify objectives and goals

The clearer your export and business goals, the better you can identify which trade show will offer the most value to your company, and the more focus your overall exhibit will have. Your goals can pertain to interested markets, theme, down to the number and type of quality leads you hope to generate.

Questions to ask include:

Is your product or service ready to enter the market?

What are your objectives in participating in a trade show?

Who is your target market and audience?

How will this trade show complement your export strategy?

Do you have the necessary resources: people, finance, time, commitment?

Do you have the capacity to follow-up on new leads?

Consult exhibitor and attendee information

Proactively contact show organizers, websites, and all other available sources well in advance of your commitment to exhibit. This ensures a well-informed decision is made regarding whether the event will be a value-added business activity for your organization. Understand your options—you may choose to exhibit or instead simply attend to gather contacts and market intelligence if you have insufficient resources.

4. Planning your participation

The effort and time dedicated to pre-show preparation may impact your success during and post show. Pre-show training sessions are often offered on-site prior to the show and during the show. This kind of preparation can benefit your level of efficiency and professionalism within a trade show setting.

➢ Set goals that compliment your marketing objectives and overall business strategy

It is important to understand that trade show participation extends beyond the few days you are exhibiting; it is a marketing tool that must be integrated into your overall business strategy and one that must be done properly in order to obtain results. Setting goals and developing a roadmap are key ways to achieve this.

For example, your company may sell 100 different products, but through goal setting you choose to focus on 10 of your products for trade show promotion and sales development. Refining the items you plan to display, sample, or discuss with potential clients gives your exhibit a stronger, more cohesive, and tailored focus.

➤ Inform clients of your participation

As you prepare to participate in your trade show, you should communicate with your current and potential clients:

Contact current business partners. Use your trade show presence as a forum to not only generate new business, but to solidify and build on current ones.

Notify prospective clients of your participation. Use your trade show presence as a tool for setting up first-time interactions with your desired target audience.

Fact: 71 percent of companies say their primary reason for exhibiting internationally is to increase leads and sales and to build relationships with clients/prospects. Source: EXHIBITOR Magazine, 2019.

To aid in pre-show preparations and ensure that you make the best use of your time at the trade show, it is a good idea to book appointments with prospective clients and buyers ahead of time if possible. Trade Commissioners in local markets and Agriculture and Agri-Food Canada Regional Offices can provide information on meetings that may be taking place with potential buyers that Canadian exporters can participate in.

➤ Designing a booth display

Trade shows require professional displays, promotional materials such as product or service samples, staffing, shipping and travel fees. It is important for an exporter to budget for these expenses in designing their booth display.

The importance of creating a fresh, innovative and unmatched display and selling technique is vital to trade show success. A booth that is not accommodating or welcoming to attendees will reduce the time they are willing to spend at your booth.

Tips:

Avoid confusing displays. Have distinct focal points as opposed to numerous competing designs. Within a few seconds a passerby will appraise each booth and decide on approaching.

Pavilions and booths are often exhibitor-friendly, but also have to be attendee friendly. Counters or free-standing displays should not block the view of backdrops or inhibit visitors from entering the booth.

Attempt to communicate the spirit and flavor of your organization. If you are participating in the Canada Pavilion, find out what the pavilion looks like and how your exhibit can complement the overall image and reputation your country is endorsing.

Use strong colors in displays that work in conjunction with the overall pavilion look and feel, but remember the eye appreciates and is drawn to less aggressive colors; seek a balance.

Regardless of whether you are exhibiting independently or as part of a larger pavilion, it is important to pay attention to detail to ensure the impression and atmosphere you are creating is conducive to successful business activity.

➤ Begin developing promotional material early

Understand the true nature of the event. Some trade shows are primarily focused on showcasing goods and services, while some are focused on actually conducting business and negotiating contracts. You need the appropriate materials in either situation.

All sales literature and marketing materials should be printed in the local language if exhibiting internationally, this emphasizes a professional image.

Product samples should be offered to qualified booth visitors with high clientele potential as opposed to every passerby. Remember, smaller sample sizes often send out a higher perceived value.

Consider promotional methods that extend beyond merely product samples and the handing out of written materials. Consider mailing product samples and written information of a larger quantity to qualified leads after the show. This addresses the importance of follow-up and ensures costly samples are being dispensed in an effective manner.

➤ Customs and import regulations

Find the best methods and approach to getting your exhibit and samples or products to the trade event location. Consider the customs and import regulations, as well as licenses needed for your sample products. Knowing these requirements in advance will ensure a smooth transition from pre-show to at-show execution.

Understanding the market access issues in the market you are interested in will also help determine where you will have the greatest chance of exporting success in the long run.

➤ Staff training

A pro-active, friendly and well-trained booth staff member is often more effective in attracting attendees than a free sample or giveaway. Sample engagement and qualifying questions should be given to staffers prior to the show. Avoid a scripted feel, but ensure there are guidelines on appropriate staff/attendee dialogue.

Fact: The majority of exhibiting companies train their staff on how to qualify leads at trade shows. Source: EXHIBITOR Magazine.

The clients you gain become your company's ambassadors. If attendees are happy with the interaction with your staff, they are more likely to think positively of your company, more willing to do business with you, and more eager to spread the word about your product or service to others.

Tips:

Develop qualifying questions for staff to use to pinpoint the audience you are looking to reach and to verify booth visitors as potential clients. Perform these types of activities before getting into product sampling or demonstration.

Do not employ booth staff that cannot answer questions. Not only should booth staff be able to answer basic questions pertaining to your product or service, they should be able to answer extensive questions pertaining to your company's capabilities, export intentions, current market exposure and efforts to expand into other markets.

Ensure booth staff are clearly identified and professionally dressed and focused on being open and approachable to attendees.

A translator may be required at your booth for attendees who may prefer to do business in their local language.

Provide staff with business cards and other promotional material that can be handed out to booth visitors if appropriate.

Never leave the booth vacant. A booth staff member must always be present.

Fact: Attendees spend on average 8.4 *hours per show visiting exhibits. Source: EXHIBITOR Magazine.*

5. Participating on-site

No matter how much preparation goes in, there will bound to be changes and surprises. Be prepared for change and be prepared to adapt.

➢ Arrival

Aim to arrive at the show at least 24 hours in advance. You may need to allow for extra time if exhibiting overseas.

When you arrive, get in touch with your local contact if you have one to stay updated on what is happening locally.

Check that you have all documentation related to shipping invoices, booth registration etc. with you.

➢ Booth set up

Ensure you obtain a floor plan and take time to familiarize yourself with the surroundings and other exhibitors.

Check that your display is set up properly.

Ensure all promotional material is ready.

Organize a final staff briefing to go over objectives, plan of action, and schedules.

Reconfirm your upcoming trade show appointments.

➢ Understanding buyers

Trade show crowds exhibit certain behavioral patterns. Taking the time to understand the audience will ensure that you develop a good approach to getting them interested in your booth, products and services. Cues can be taken from the type of workshops or seminars organized in conjunction with the show, local events, as well as the local/regional culture. Trade Commissioners and regional office staff can also provide advice.

Your staff have to understand your organization's business plan and target audience and not just push a product. Their efforts will be lost on attendees if they are not able to transfer important product and company information as well as provide a professional and competent representation to prospective clients.

➢ Spotting genuine leads

The key to garnering successful leads is being able to recognize a genuine prospect. Building a profile of your target audience beforehand and being aware of whether you are seeking local partnerships or international business alliances can aid staff in identifying the right leads. Developing qualifying questions beforehand can also help staff.

Fact: 95 *percent of companies say lead acquisition is one of their most important exhibit-marketing objectives. Source: Center for Exhibition Industry Research,* 2018.

Clear business goals will assist in the identification of qualified booth visitors. For example, if you have decided to focus on business expansion into the United States, then all visitors to your booth from the United States have already met one important qualifying variable. The more variables met, the greater the significance of product samples and information.

➤ Networking tips

Be prepared to have your company information and contact details on hand, whether it is using business cards, pass scanners or Quick Response (QR) codes.

Be proactive in approaching prospects. Introduce yourself. Develop an elevator pitch or opener that highlights your business and product.

If your staff need to sit, provide stools rather than chairs so that staff members can greet booth visitors at eye level.

Avoid closed questions-you want to understand what your prospective clients are looking for and what has caught their attention at your booth.

Ensure booth staff know your key messages. Active listening is important and booth staff should pay attention to body language, use of words, and tone of voice.

If you are conducting a demonstration, spot individuals who express above average interest in your products or services. Get them involved in the demonstration or ensure you reach out to them afterwards.

Fact: Trade shows attract attendees who have a major say in purchasing decisions. Half of all attendees have a buying plan in place before their visit. Source: Center for Exhibition Industry Research.

➤ Lead management

Your exhibit should target specific prospects that are interested in the products you are selling. Once targeted, the next step will be to initiate person-to-person contact. A system should be developed to record down the contacts made. Capturing the leads generated can be done by filing away business cards received or noting down each interaction in writing or electronically.

Tips and best practices:

Ensure there is a lead management system in place during the show. Staff can break down leads into several categories, such as high, medium and low interest/potential to prioritize who to follow up with.

Decide on a common method of follow-up prior to a show for staff to follow and allocate sufficient time and resources for post-show lead follow-up.

Fact: Nearly half (49 *percent) of exhibitors are able to track what percentage of leads sourced at a given show ultimately convert to sales. Source: EXHIBITOR Magazine,* 2020.

6. Post-show and measuring performance

After a trade show ends, it is easy to fall behind on lead follow-up. In most cases trade shows are about generating qualified leads and not about conducting final sales. The longer one leaves leads unattended, the less likely they are to evolve into business opportunities.

➤ Within one month

Professional follow-up within a recommended 30-day period helps to ensure your business reaps the most benefit from trade show participation; otherwise, the effort and time invested in participation can be lost.

Initial follow-up can be conducted in a variety of ways. It can be made by a simple phone call within a few days of the show, while more intensive follow-up should be made within the span of a month (timeframes can vary depending on the nature and seasonality of your business). Examples of follow-up include thank you letters or e-mails, trips to visit leads, and distribution of product samples and information packages.

You may also decide to extend your stay following a trade show to conduct immediate face-to-face follow-up with serious leads.

➤ From six months to one year

Stay in touch with your leads by keeping them informed about new and upcoming products and services, as well as updates on existing ones. You may offer a newsletter for subscription or develop social media content to disseminate your company's latest news.

Are you thinking of attending or exhibiting at other trade shows? As your company plans ahead for future trade show opportunities, let your leads know which upcoming trade shows you are looking to attend, as they may be considering attending too, providing additional opportunities to connect.

➤ Performance evaluation

However formal or informal, ensure an event debrief and show evaluation is conducted analyzing what worked, what did not, and what could be improved. It is best to hold a debrief soon after the event has concluded while the experience is still fresh in your mind.

Feedback can be useful to ensure that each trade show experience is an improvement from the last. It can also serve as the basis for selecting the "right" shows to attend in the future. The quality of the leads generated, sales revenue generated, as well as visitor traffic are examples of what can be measured to get a picture of the overall success of your booth.

A post-show report covering show description, market opportunities, new trends, lessons learned and recommendations for future trade shows can be useful for future event budgeting and planning. In addition, a debrief on your marketing strategy and its impact and correlation to results achieved can aid in crafting future marketing plans.

Use your trade show participation, however frequent, as an opportunity to consistently refine your approach to business within a trade show setting. Ensure it satisfies your business needs, fulfills your business goals and provides some form of return on your time and financial investments.

There is no definitive approach to trade show participation, only suggestions as to how to better approach this valuable marketing method. It is important to customize your approach to exhibiting to maximize your success.

 NEW WORDS AND PHRASES

interact   vi. 交流; 沟通; 合作

forge   v. 努力地缔造

incorporate　vt. 使并入

forum　n. 论坛

profile　n. 形象

venue　n. 举办场所

cost-effective　adj. 有最佳利润的; 有成本效益的; 划算的

marketing strategy　营销策略

lead up to sth.　是……的先导; 是导致……的原因

in place　准备妥当

convert　v. 可转变为

target audience　目标客户; 目标受众

tailored　adj. 特制的; 专门的

budget　n. 预算

mindful　adj. 想着; 考虑到

incur　vt. 引致, 带来(成本、花费等)

foreign exchange rate　外汇汇率

marketing tool　营销工具

sponsorship　n. 资助

boost　vt. 使增长; 使兴旺

real-time　adj. 实时的

leveraging social media　利用社交媒体

innovative　adj. 创新的

differentiate　v. 区分; 区别

prominently　adv. 突出地

strategically　adv. 战略上

access　vt. 访问; 使用

extended sale force　扩大销售队伍

implement　vt. 执行; 实施

evaluate　vt. 评估

execution　n. 执行; 实施

timeline　n. 限期

on track　稳步前进

assessment　n. 评估

align　v. 使一致

Trade Commissioner　贸易专员

Agriculture and Agri-Food Canada (AAFC)　加拿大农业及农业食品部

Regional Office Expertise　区域办事处专业知识

sector　n. 部门

stakeholder　n. 参与方

pertain to sth./sb.　与……相关; 关于

proactive　adj. 积极主动的

option　n. 选择

pre-show preparation　展前准备工作

pre-show training session　展前培训

prior to　在前面的

cohesive　adj. 有聚合力的

solidify　v. 巩固

appraise　vt. 评价；估量

pavilion　n. 展览馆

endorse　vt. 支持；认可

showcast　vt. 展示

clientele　n. 客户；顾客

ambassador　n. 大使

pinpoint　vt. 明确指出；确定

verify　vt. 核实

shipping invoice　货运发票

booth registration　展位登记

familiarize　vt. 使熟悉

promotional material　宣传材料

seminar　n. 研讨会

garner　vt. 获得；收集

alliance　n. 同盟，联盟

scanner　n. 扫描仪

Quick Response code　二维码

highlight　vt. 突出；强调

person-to-person contact　人际交往

record down　记录下来

capture　vt. 捕获

lead management system　客户线索管理系统

prioritize　v. 优先考虑

timeframe　n. 时间表

seasonality　n. 季节性

upcoming　adj. 即将发生的

subscription　n. 订阅

disseminate　vt. 传播

debrief　n. 汇报

sales revenue　销售收入

correlation　n. 相互关系

refine　vt. 改进；改善

customize　vt. 个性化设置

NOTES

1. Facebook（脸书）公司创立于 2004 年 2 月 4 日，总部位于美国加利福尼亚州门洛帕克。其主要创始人为马克·扎克伯格（Mark Zuckerberg）。2012 年 3 月 6 日发布 Windows 版桌面聊天软件 Facebook Messenger。2021 年 Facebook 将公司名称改为 Meta，表示其重大品牌重塑计划的一部分。

2. Twitter（推特）是一家美国社交网络及微博客服务的网站，是全球互联网上访问量最大的十个网站之一，类似于中国的微博。

3. Instagram（照片墙）是一款运行在移动端上的社交应用，其可以快速、有趣的方式将你随时抓拍下的图片在应用上进行分享。

4. Linkedin（领英）是一个面向职场的社交平台，总部设在美国加利福尼亚州的森尼韦尔。该公司于 2011 年 5 月 20 日在纽约证券交易所上市。网站的目的是让注册用户维护他们在商业交往中认识并信任的联系人，俗称"人脉"。用户可以邀请他认识的人成为关系（Connections）圈的人。

5. Flickr 是雅虎旗下图片分享网站。其是一家提供免费及付费数位照片储存、分享方案的线上服务与网络社群服务的平台。其重要特点是基于社会网络的人际关系拓展与内容组织。这个网站的功能很强大，已超出了一般的图片服务，比如图片服务、联系人服务、组群服务。

6. YouTube 是全球视频播客类网站的翘楚，可供网民下载、观看、分享影片或短片，是全世界访问量最大的视频网站。

7. Elevator Pitch（电梯法则），也被称为电梯游说，即用极具吸引力的方式简明扼要地阐述自己的观点。例如你在电梯里，只有 30 秒的时间来向一位关系公司前途的大客户推广产品且必须成功。

8. Market Exposure/Market Risk，是指未来市场价格（利率、汇票、股票价格和商品价格）的不确定性对企业实现其既定目标的不利影响。

EXERCISES

Ⅰ. Answer the following questions.

1. How do you build your marketing strategy?

2. When companies choose shows, what are your suggestions?

3. What a company should do prior to a trade show?

4. List the preparations of participating on-site trade show.

5. Why is professional follow-up within one month recommended? and what can be done?

Ⅱ. Please translate the following English sentences into Chinese.

1. Trade shows allow companies to interact and forge new partnerships with a diverse group of potential buyers and clients, all within one location.

2. Ensure that the social media tools for sharing and embedding your content are displayed prominently, strategically located, and easy to access whether it is on promotional material or on your website.

3. The clearer your export and business goals, the better you can identify which trade show

will offer the most value to your company, and the more focus your overall exhibit will have.

4. It is a marketing tool that must be integrated into your overall business strategy and one that must be done properly in order to obtain results.

5. 71 percent of companies say their primary reason for exhibiting internationally is to increase leads and sales and to build relationships with clients/prospects.

6. To aid in pre-show preparations and ensure that you make the best use of your time at the trade show, it is a good idea to book appointments with prospective clients and buyers ahead of time if possible.

7. The importance of creating a fresh, innovative and unmatched display and selling technique is vital to trade show success. A booth that is not accommodating or welcoming to attendees will reduce the time they are willing to spend at your booth.

8. Regardless of whether you are exhibiting independently or as part of a larger pavilion, it is important to pay attention to detail to ensure the impression and atmosphere you are creating is conducive to successful business activity.

9. Not only should booth staff be able to answer basic questions pertaining to your product or service, they should be able to answer extensive questions pertaining to your company's capabilities, export intentions, current market exposure and efforts to expand into other markets.

10. Their efforts will be lost on attendees if they are not able to transfer important product and company information as well as provide a professional and competent representation to prospective clients.

Ⅲ. There are 10 sentences in this section. Beneath each sentence there are four words or phrases marked A, B, C and D. Choose one word or phrase that best completes the sentence.

1. Once you have decided to participate and exhibit at a trade show, it is very important to create a (   ) to promote your brand both leading up to and during the event.

A. marketing mix   B. marketing strategy
C. price mix   D. promotion strategy

2. (   ) can boost your trade show strategy by providing real-time client interaction and tailored content for your target audience.

A. Advertising   B. Public relations
C. Individual appointments   D. Effective use of social media

3. From the perspective of booth marketing strategy, booth marketing strategy is the scientific configuration and effective combination of seven elements such as product, price, place, promotion, people, physical evidence and process, namely (   ).

A. 7Ss marketing strategy   B. 7Ps marketing strategy
C. 7Ps promotion strategy   D. 7Ps pricing strategy

4. Before the opening of the exhibition, exhibitors should generally prepare sales literature and marketing materials, among which product samples should be offered to (   ).

A. exhibitors   B. organizers
C. qualified booth visitors   D. ordinary viewers

5. Develop (   ) for staff to use to pinpoint the audience you are looking to reach and to verify booth visitors as potential clients.

A. qualifying questions   B. sampling

C. demonstration   D. sales literature

6. (   ) is the closest in meaning with the phrase "trade commissioner"?

A. trade intermediary   B. trade dealer

C. trade assistant   D. trade exhibitor

7. Your exhibit should target specific prospects that are interested in the products you are selling. Once targeted, the next step will be to initiate (   ).

A. business-to-business contact   B. business-to-person contact

C. leaders-to-leader contact   D. person-to-person contact

8. The phrase "elevator pitch" can be translated into the following ones EXCEPT (   )?

A. 电梯法则   B. 电梯游说   C. 电梯陈述   D. 电梯高点

9. Professional follow-up within a recommended (   ) period helps to ensure your business reaps the most benefit from trade show participation.

A. 15-day   B. 30-day   C. 5-day   D. 60-day

10. (   ) covers show description, market opportunities, new trends, lessons learned and recommendations for future trade shows.

A. Feedback   B. A debrief

C. A post-show report   D. Performance evaluation

Ⅳ. Writing

Letter for Visa application

Mr. Blare Brown, vice president of US MICE Consulting LLC. who is invited to China by QQQ Technology Co., Ltd to attend Canton Fair. As secretary of QQQ company, you are required to write a letter to Mr. Blare Brown for the assistance of his visa application.

During Mr. Blare's stay in China (10th April, 2023—19th April, 2023), he will visit your stand and have a business talk with you company's president.

Other information needed in the letter could be contrived as you like.

📖 Reading 2

Preparing Sales Literature and Exhibit

Sales literature refers to all the written material that enables an enterprise to inform customers and potential buyers about the existence of a product or service and its characteristics. It simplifies the buying process and once a product is bought, sales literature helps the person use the product. Sales literature is used to support the marketing strategy of an enterprise and back its communication capabilities.

All things that do not simplify the buying process, or provide instructions on how to use a product are not considered as sales literature. Things like receipts or letters from the enterprise thanking a customer are not sales literature.

Sales literature must explain the benefits of the products in a clear way. In addition, it should make products look attractive and show the enterprise in a favorable light. As with all elements of the marketing communication mix, literature needs to be carefully designed, well printed and free from mistakes, if a good impression is to be given. To prepare sales literature the enterprise should rely on personnel with adequate time, background and ability.

No less important, it must somehow be put in the hands of the people who should read it. The enterprise's sales literature must reach the target markets. Distribution by direct mail, by salesmen making personal calls, and at fairs and exhibitions are among most effective ways of distributing sales literature.

The most important types of sales literature are: price lists, catalogues, technical papers, instructions and brochures.

Price Lists

A price list is a comprehensive record of the products offered by the enterprise together with their prices. The price list should be aimed at a particular market segment. The enterprise should use price lists despite of fears of competition, transparency, etc.

A price list should include the following points: item code, item description, unit price, special prices, minimum order, price of complementary items, packaging costs, currency, expiration date, incoterms, average delivery time and warnings.

A price list is a marketing tool, not just a sheet of information. This means that making sure that accurate information is communicated is as important as a good presentation. The price list should be nicely formatted, it should include the company logo and colors, and it should be used to make the product look attractive to the potential buyers.

Before its distribution the price list should be verified from the production and the marketing points of view to make sure that prices are adequate to guarantee the long-term profitability of the business. The price list is a valuable source of information for the competition. Select carefully the people this information will be distributed to.

Catalogues

A catalogue is a publication that shows a variety of products offered by a company or retailer to the market. The enterprise needs catalogues to sell its products to current and potential buyers.

Before starting a catalogue it is necessary to know who the client is. Determining whom the catalogue aims to address is important because a single catalogue could hardly be appropriate to target business-to-business, retailers and final consumers simultaneously. Enterprises can have more than one business; in this case it is also advisable to design specific catalogues for each business and targeted segment.

The catalogue should include basic information such as contact details, a presentation of the enterprise, a brief history of the company and its vision, the business mission, the values of the enterprise, information on awards and honors received, certifications, item references, physical dimensions of each product (note that the units used are sensitive to regions) and weight. It is also useful to include information such as how much of the product fits in a container and even information on transport costs.

Photographs and graphic designs are one of the most important parts of the catalogue. They serve the purpose of attracting the client as well as making him compelled to read it. On this matter it is advisable to obtain professional expertise. Showing real situations with people using the products can make the offer more appealing. The pictures can include the accessories of the products when it is in the interest of the business. It is useful to group products by collections or lines.

When making a catalogue one should keep in mind that it is the enterprise's business card, i. e. a way of presenting the enterprise to potential clients. Take care of every detail. Make it attractive. The manager should look at his enterprise's catalogue through the eyes of the client. Make it simple to use, do not overdo it by including more than the necessary information. Ask yourself the basic questions that a customer would ask when looking at the catalogue.

Some effective means of distributing catalogues are: direct shipments by mail to potential clients, trade fairs and similar events, export promotion organizations, chambers of commerce, etc. The catalogue should be written in several languages when used for export.

Catalogues are usually updated on a yearly basis. If the enterprise makes new products more frequently the catalogues should be updated accordingly. Perhaps new pages can be added or new updated information can be sent. The catalogue should contain the date and validity of the prices and the information included. In addition, it should be simple to use for the customer.

Technical Papers

Technical papers provide additional relevant specialized information on the products. Technical information can enhance the product, making the purchasing decision easier.

The main types of technical papers are: technical specifications, terms of sale and guarantees.

Technical specifications include: description of the raw material used, description of the product properties (i. e. elasticity, resistance or other chemical or physical properties relevant to the use of the product), description and photographs of different possible uses, descriptions and photographs of the different packaging options.

The terms of sale help customers to place orders because they describe all the arrangements. They can include information on: customer service, order procedures, conditions of payment, special orders, refunds and guarantees.

Guarantees are a common feature offered to the clients in order to fulfil legal requirements. In addition, they can give the enterprise a competitive edge against others companies. In any case, the enterprise must be committed to keeping the promise and good attitude in order to achieve a total customer satisfaction. A guarantee should have the following elements: time frame of validity, receipt of purchase as a proof or evidence that the item was purchased and when, what it covers, what it does not cover, procedures to follow, and whether the guarantee is for personal or commercial use.

To assess whether technical papers are needed the type of product and the targeted segment should be analyzed. When the product has a high level of complexity technical papers are needed. The analysis of the segment will determine whether the market demands some more information other than catalogues and price lists.

The most efficient way to distribute technical papers is to send them along with the catalogues. When writing technical papers, it should be kept in mind that the presentation can play an important role in support of the strategy. No one likes reading excessively technical literature. Keep a balance, and most importantly, make it understandable.

Instructions

Instructions are directions that tell a person how to assemble, maintain or use an item or product. They are used to let the person know how something works, making sure information is provided on maintenance and assembly.

There are four types of instructions: assembly instructions, usage instructions, care and maintenance instructions and hazards and security instructions. Instructions on assembly are needed when the product is sold in pieces that need to be put together. In order to make them easier for the person to understand it is recommendable to include pictures or graphs explaining the assembly process. Always provide instructions on how to use the product. Photos and examples can be very helpful for this matter. Care and maintenance instructions allow users to know how to take care of the product and where to obtain maintenance services. Some products must specify the dangers and hazards to a person's health.

The most efficient way to distribute the instructions is together with each product. A proper and clear set of instructions can be attached to the product itself or to the packaging. Things like household appliances have separate sheets of paper or small booklets with instructions.

Before attaching the instructions to the product make sure they are clearly written and are easy to read. A way to check that instructions are well done is to have a person who knows little about the product read them. This person should be able to assemble a toy, for example, just by reading the instructions. It may be necessary to write the instructions in several languages. Verify whether the targeted market has specific regulations or instructions for hazardous materials.

Brochures

A brochure is a type of small magazine that contains pictures and information on a product or a company.

The main types of brochures are: product or service brochures, and corporate brochures. The first type of brochures emphasizes special features of specific products or services and the way these satisfy the buyer's needs. Corporate brochures focus on the company and provide relevant information on its different businesses as well as information on corporate values, achievements, etc.

Most small businesses use brochures to list the products or services they provide together with their credentials for providing them. One of the most important elements on the cover is the company name and contact details.

Graphical elements and stylistic features play an important role to communicate the message in a limited space. As in any other design project, contrast adds visual interest to a page and helps to highlight the most important points. Managers should use contrast in the typefaces, rules, colors, spacing, size of elements, etc. Repeat various elements, such as colors, typefaces, rules, spatial arrangements, bullets, etc., in the design to create a unified look. Strong, sharp edges create a strong, sharp impression. A combination of alignments (using centered, flush left, and flush right in one piece) usually creates a sloppy, weak impression. The design principle of proximity, or grouping similar items close together, is especially important in a project such as a brochure where there is a variety of subtopics within one main topic. How close and how far away items are from each other communicates the relationships among the items.

The content of the brochure should be kept as short as possible. The shorter the paragraph, the more likely it will be read. Limited bullets, good use of color, and lots of white space—all these elements enhance the company's message by making the copy easy to read. A few strong, brief points are far more effective than dozens of weak ones.

Unsolicited brochures are rarely read. Therefore, one of the most efficient ways to distribute them is at fairs and exhibitions. Other ways of distributing brochures are through salesmen, displaying them at the point of sale, or attaching them to an order made by a client.

(Edited from The Tasks of Business Management System by ITC)

 ## NEW WORDS AND PHRASES

sale literature　销售材料; 销售说明书

receipt　n. 收据

transparency　n. 透明度

minimum order　最低订购量

currency　n. 货币

incoterms　n. 国际贸易术语解释通则

International Chamber of Commerce　国际商会

logo　n. 商标; 标识语

adequate　adj. 充足的

retailer　n. 零售商

business-to-business　n. 企业对企业电子商务

simultaneously　adv. 同时地

segment　n. 段; 部分

certification　n. 证明

dimension　n. 尺寸

container　n. 集装箱; 容器

graphic　adj. 形象的; 图表的

compel　vt. 强迫, 迫使

appealing　adj. 吸引人的; 动人的; 引起兴趣的

accessories　n. 附件(accessory 的复数形式)

overdo　vt. 把……做得过分

chamber of commerce　商会

specialized　adj. 专业的; 专门的

technical specification　技术规范; 技术说明

legal　adj. 法律的; 合法的; 法定的

be committed to　致力于

validity　n. 有效性

purchase　n. 购买　vt. 购买

complexity　n. 复杂性

assemble　v. 装配

maintain　v. 保养

maintenance　n. 维护, 维修

assembly　n. 装配

hazard　n. 危险

recommendable　adj. 可推荐的; 值得推荐的

specify　vt. 详细说明

distribute　vt. 分发

appliance　n. 器具; 装置

regulation　n. 规则

brochure　n. 手册, 小册子

credential　n. 证书; 凭据

element　n. 元素; 要素

graphical　adj. 用图表示的

bullet　n. 项目符号

alignment　n. 成直线

sloppy　adj. 草率的; 粗心的

subtopic　n. 副主题

unsolicited　adj. 未经请求的

 NOTES

1. Market segment（细分市场）是指将消费者按照不同的需求、特征区分成若干个不同的群体，而形成的不同的消费群。

2. Chambers of commerce（商会）。China Chamber of International Commerce（中国国际商会）是由设立于中国、从事国际性商业活动的企业及与之相关的组织组成的全国性社会团体，其业务主管单位是中国国际贸易促进委员会（简称中国贸促会）。

3. Sales literature（销售资料）是指使企业能够将产品或服务的存在及其特点告知顾客和潜在购买者的所有书面材料。

4. Price lists（价格表）是对企业所提供的产品及其价格的全面记录。

5. Catalogues（目录）是一个公司或零售商向市场提供的各种产品的出版物。

6. Technical papers（技术文件）提供有关产品的额外相关专业信息。

7. Instructions（说明书）是告诉一个人如何组装、维护或使用一件物品或产品的指示。

8. Brochures（宣传册）是一种小型杂志，包含产品或公司的图片和信息。

 **Reading 3**

Effective Advertising for Your Trade Show Booth

Business success is based on consumer knowledge. If a business is unknown to customers, it cannot become successful. Customers will not seek out your business. They will gravitate toward those that are familiar, visible, and readily available to them. Get out there and make your business known! How do you do that? Trade show marketing. Traditional advertising and promotions are still a vital function for the success of any business. There are many ways in which you can promote and advertise your brand. Participating in trade shows is one function that is a particularly effective method of promoting your business.

Whether your appearance at a trade show is a success or not depends heavily on how you advertise it. The rules for effective trade show booth marketing have changed fundamentally over the past years. Particularly social media has been taking on an ever-more important role. We show you how advertising for your booth can lead to success!

I was recently interviewed on a podcast that will be broadcast to thousands of HR professionals right before their biggest conference and trade show of the year. I am actually attending the show with a client of ours who will be exhibiting, and if you know anything about conferences, you probably know that often exhibitors are viewed as a "necessary evil". While this is largely an unfair and unfounded misnomer, there is a reason why brands and exhibitors can come across this way.

It's because of the way they often market and promote themselves before, during, and after the trade show.

The reason that I was asked to be on the podcast came down to one thing—how I had approached the marketing aspect for my client. I didn't go in there with my vendor hat on; I went in there with my marketer hat on and, therefore, was the only vendor (out of more than 600) who received such an invite. That says something friends, and I was beyond honored and proud.

Too many companies pull the "it's all about us!" card and sell, sell, sell. Look, I get it—you've spent a lot of money to exhibit at these trade shows, and you know that you had better show some ROI on that money, or someone up the corporate food chain is not going to be happy. There's a lot of pressure to stand out from all the other vendors who are there for the exact same reason you are. But what if I told you there's a way to do it, and in a way where you get your point across, drive people to your booth, and let the attendees know you have something they need.

Pre-Show Marketing

Planning—Repeat after me: you must have a plan of action in place. What is the golden action you want people to take? Stop by your booth, right? So, decide how you're going to do that. Are you going to have a giveaway? Will that giveaway be BIG like an iPad or a normal tchotchke that everyone else is giving out? Whatever it is, have that decided and go from there.

Pre-Show Mailer—I've often said that direct mail is not dead, and in this day and time, all of us are being beaten over the head with social content. A well-done direct mail piece can work really well in this instance. Most shows will offer mailing lists for sale to their exhibitors—take advantage of it. I personally prefer pre-show lists and mailers because you can target your message and direct attendees to your booth. Post-show lists don't do as much for me because you're targeting everyone who attended, and that's a wide net to cast. If the list is HUGE, cull it down to just those who are your target audience, and that will save you money on the cost of printing and postage.

Pre-Show Branding-Create new cover images for all your social media sites to let people know that you're going to be at the trade show and which booth you'll be in. Just don't forget to swap the cover images out after the show is over. Some larger conferences will even have custom exhibitor badges you can incorporate into your design.

Pre-Show social media-I spend a LOT of time looking at who's speaking and attending conferences, and I will create a special Twitter list specific to the show I'm going to attend. This does a couple of things for me. It allows me to start some dialogue with people who are involved (and IF it's natural conversation), I learn more about what the attendees are interested in, and it gives me a stream to follow while I'm at the actual event. Look to see if there is a group of "ambassadors" or bloggers who are devoted to promoting the conference, and if your company has a blog, reach out to them and ask if they would be willing to meet you for a chat and interview. This is a great opportunity to get content for YOUR blog and site as well as a way to promote them! Is your competition going? Make sure you're following along what they're doing pre-show. And of course... see if there is a conference hashtag and incorporate it when you can. I also recommend that you have your own hashtag and pull that into the mix as well.

Pre-Show Content-Look, you shouldn't be "just" selling (remember the 90/10 rule...) on social media and on your blog anyhow, but if you are, please do not increase it before a trade show. Sure, you can push out some content that lets people know that you'll be at ABC Conference, but that's enough. Instead, customize your messages as to WHY they should come by your booth while at the conference, and then use the conference hashtag in the message. This is a much, much better approach. Same thing goes for your blog—can you create content that would appeal to the attendees of the conference? Then you're actually offering them something of value, and they'll be more receptive to actually checking out what it is that you can offer them!

At the Show

Now that you're there, if you've done a great job at planning, it should be smooth sailing, and it'stime to let your team do their work!

Stay Connected-Use that Twitter list that you created and start to follow what people are saying and when appropriate, join the conversation. People on social media at a conference LOVE interacting with other people there. It's like a big family reunion. I always create a separate stream for the show hashtag, but be warned—there will be a TON of action going on, so don't feel bad if you miss some things. You are there to work the booth, remember?

Live EVERYTHING! —I'm talking live Tweeting, live streaming on Periscope, blab some interviews, Facebook Live, Instagram, Snapchat—do it all. Believe it or not, there are a lot of companies who still don't do this, and it is the PERFECT opportunity to stand out from the crowd. It will be a lot of work, but trust me, it will pay off. (Pro tip: I will sometimes throw out a "disclaimer" tweet or message just to let people know that I am on-site at an event, so my activity will be much more than a typical day.)

Be Engaging! —Yes, you need to sell your products and services, but make it fun! Let your personality shine through and get to know the people coming by the booth. I have a firm rule when it comes to working a booth—no chairs. Or rather, chairs are for guests/attendees (not for staff). Do not sit down in your booth if it's a time when the exhibit hall is actually open. If there's down-time while sessions are going on, sitting down is fine but only then. There is nothing more off-putting than walking up to a booth and seeing that. And don't even get me started about having staff sitting behind a table. Quite often, I will have it immediately removed, or at the very least, move it so that it's against the drape.

Collect Leads and Information—Whether you're using a lead generation app/scanner that's conference provided or collecting business cards, don't make the rookie mistake of getting nothing. This is how I follow up after a show (instead of buying the post-show mailer), since these are actually people who expressed an interest. If you have a really good conversation with someone (aka a hot lead), make sure you note that on his/her card or somewhere. I even put them in a special place separate from the others, since I know that they need to be followed up with ASAP after the show.

Post-Show Marketing

Follow-Up-Whether you are using the leads you collected on your own, or you decide to buy the post-show list, have a plan in place to send out your piece (whether it's electronic or direct mail) the week after the show ends. Another downside to post-show lists is that everyone is following the same timeline, so your message is going to be one of many. I find that personal emails to the top people you met works best. Remind them of the conversation, and set up a time to continue the conversation.

In closing, the biggest thing to keep in mind when it comes to marketing yourself before a conference or trade show is to have a plan, and make sure your message gives attendees a clear and compelling reason to take the time to see you. Use simple, concise words that can easily be consumed quickly. It also pays to remember that they have an agenda and very full schedule aswell.

 NEW WORDS AND PHRASES

seek out   挑选

gravitate toward   被吸引到

trade show marketing   展会营销

fundamentally   adv. 根本上; 完全地

take on a role   承担角色

podcast   n. 播客

view as   把……视为

necessary evil   一件不想做但必须做的事情

largely   adv. 很大程度上, 主要地

unfounded   adj. 莫须有的; 不依据事实的

misnomer   n. 用词不当

come across 被理解

come down to 可归结为

vendor n. 小贩, 摊贩

invite n. 邀请; 请柬

giveaway n. 赠品

tchotchke n. 小玩意儿, 不值钱的小摆设

give out 分发

dead adj. 过时的, 不再重要的

social content 社交内容

mailing list 邮寄名单

cull vt. 挑选

swap v. 交换, 替换

custom exhibitor badge 定制参展商标记

blogger n. 写博客的人

hashtag n. 标签

appeal to 引起兴趣

check out 核实

Tweeting 在社交网站 Twitter 上发微博

blab v. 泄露

pro tip 专业建议

throw out 随口说; 脱口而出

disclaimer n. 免责声明

off-putting adj. 令人讨厌的; 让人绕着走的

drape n. 帷幕

rookie n. 新手

aka abbr. (also known as) 亦称为

ASAP (as soon as possible) 尽早

downside n. 缺点, 不利方面

NOTES

1. ROI（投资回报率）是指通过投资而应返回的价值，即企业从一项投资活动中得到的经济回报。它涵盖了企业的获利目标。利润和投入经营所必备的财产相关，因为管理人员必须通过投资和现有财产获得利润。投资可分为实业投资和金融投资两大类，平常所说的金融投资主要是指证券投资。

2. Periscope，流媒体直播服务运营商。Twitter 于 2015 年 3 月以接近 1 亿美元的价格收购了提供流媒体直播服务的 Periscope。4 月初，Periscope 正式上线。

3. Snapchat（色拉布）是由斯坦福大学两位学生开发的一款"阅后即焚"照片分享应用。利用该应用程序，用户可以拍照、录制视频、添加文字和图画，并将它们发送到自己在该应用上的好友列表。这些照片及视频被称为"快照"（snaps），而该软件的用户自称为"快照族"（snubs）。

ENGLISH FOR WORKPLACE COMMUNICATION

Sample Dialogue: Trade fair registration

Situation: The following conversation is between Conference Assistant (CA), John Smith, the Conference Manager (CM), Tony (T), and May (M). May and Tony go together and visit the booth in a show. They go to the exhibition registration office to check in.

CA: Good morning. Can I help you?

M: Yes, good morning. Er, well, we've come to register for the conference. I'm May Hunter and this is my colleague, Tony Marshall.

CA: Ah, yes. Here are your conference badges and this is your information pack.

T: Thanks. Is there somewhere we can get some coffee?

CA: Of course. Now, you're with "Comfort Tours", so your stand is number 35, over there, right by the coffee shop. It should be open by now.

T: Thanks, that's great. That's a good place to have the stand, May, lots of people will pass by, and we can always pop across for a coffee ourselves!

M: Can we set up the stand now?

CA: Yes, whenever you wish.

M: Oh look, there's the Conference manager. Shall we say hello? Excuse me...

CM: Oh, hello. Pleased to meet you. I'm John Smith, the Conference Manager.

M: Pleased to meet you.

CM: If you need anything, just ask me or one of my staff.

T: Thank you. Well, this is lucky. It looks as if the exhibition is well-organized. I think we are going to have a good time here!

M: I suppose we should contact Head Office to report that we've got here, OK?

T: Well... can't we do that tomorrow? Let's go and have a coffee!

M: OK. But after we have set up the stand, Tony! And then we can go and look at the other stands and the conference hall.

T: That's a deal, May; stand first and then some coffee.

UNIT 3

ATTENDING A TRADE FAIR

本章导读

　　参加展览会是获取目标客户和市场资讯最直接的方式，因此对于参展商来讲尤为重要。参展不仅是卖出商品，更多的是要通过参展发现自己产品的不足和差距，找到更好的解决方案。参展过程还是一个学习的过程，要始终抱着一种学习的态度，学同行、学客户、学产品、学知识。

　　通过本章学习深入理解如何在展览会上获取目标客户。作为工作人员，你在展览会上的一举一动都会影响参展效果，因此，参展商要得到客户的信任应具备以下素质：有计划，有目标，有策略，有智慧，有韧劲。本章三篇文章从不同角度讲述了参展获客的重要性以及如何获客，以提升参展的回报率。

知识精讲

展会知识

给我两个月，订单没问题！

　　这几天，全国各地陆续复工复产，史上最长的"春节假期"正式画上了句号。而在万里之外的欧洲，有一批中国员工却从除夕到现在一直没停工。

　　每年1—3月的全球展会旺季，这些来自中国的出口制造企业代表都会赴欧参展。这不，在法兰克福春季国际消费品展上，岛叔也跟这些中小企业主好好聊了聊。

一

　　崔翔勇来自河北的一家会展公司，春节期间忙着参加展会已是家常便饭。但今年这遭遇还是头一回。

　　"3号从石家庄出发，一路被测了8次体温，填了3回表格"，崔翔勇掰着指头跟岛叔算。

　　"那我们测的次数比这个还多，都数不清了。"旁边展位的吕老板听在耳里也附和道。

　　吕老板来自浙江永康，经营着一家有七八十个员工的铸铁壶厂。春节期间，他自永康前往义乌机场，因当时义乌禁止外地车辆进入，经过层层检查、测温登记后，他才被当地出租车送至机场。

受疫情影响，法兰克福官方原本预计至少15%的中国展商会缺席，但最终结果显示，报名参展的664家中国企业，只有7%没来。

"法兰克福消费品展有170多个国家和地区的采购商参与，是我们这行最重要的展会，无论如何都得来。"邢台三厦集团的总经理韩光辉聊起坚持参展的决心如是说。

他所在的企业主要做铸铁锅，德国某著名厨具品牌的产品均由其代工，一年出口额能达到7 000万美元。

"外贸这行的惯例是'见样下单'，互联网再发达，你也不能光用图片做生意，一定得见面。对中国中小企业来说，参加展会花钱最少，效果最好。"崔翔勇说。

据他介绍，企业主和外国客户一般是每年2月在法兰克福碰面，中间经过两个月洽谈，4月份客户到广交会再见面，然后参观工厂，订单基本就确认了。

"如果这次不来，后面的事就都没了。"

二

"那个坐头等舱过来的人"——这是单磊最近的绰号。

刚听到这一称呼，岛叔脑子里浮现出一个中国大老板的形象；等走到只有5平方米大小、摆满刀叉的展位时，眼前所见却是一位憨厚的山东大哥："实在没办法了，一张头等舱要价27 000元，原定的3个人就来了我一个。"

单磊做不锈钢餐具已有30年，70%的订单出口到欧洲。今年，他早早订下青岛经北京到欧洲的航班，疫情爆发后，他把航班改成了直飞莫斯科再转机："当时想着减少国内乘机次数，没想到反而更麻烦了。"

2月2日上午从青岛起飞4小时后，空姐通知"飞机即将降落青岛"；单磊以为空姐在跟乘客"逗闷子"，谁知该航班已被俄方禁止降落，兜了一圈又回到原地。

"当时就买了头等舱改飞北京，从北京去欧洲。我们跟客户打过招呼，一定准时到。"单磊的客户大都是合作十年以上的老客户，每年按时见面已成惯例。

"除非航班一个都没有，不然我不会迟到。"他说。

几经辗转飞抵法兰克福时，16箱样品已安放于展馆，往年3个人，一上午就能布好展位，这次单磊一个人忙到了快闭馆。

据他回忆，客户也有顾虑，还问能不能把见面地点换到亚洲馆外面，"不过戴口罩的客户是少数，而且到展台就把口罩摘了，还道歉说知道这样不太礼貌"。

"这咱都能理解。但人家来你展位，就是想做生意、支持你的。"正跟岛叔聊天的时候，一个拉美老客户来到单磊展位，二人约着一起吃晚餐。

韩光辉每次来法兰克福，也总要跟德国大客户吃饭，今年赶上疫情，他事先还询问了一句："是不是就不吃饭了？"

结果对方说："为什么不吃？"

"最后跟往年一样，喝着啤酒红酒，餐桌上聊得很好。"韩光辉开心地说。

三

"参展第一天只收到一张名片，但第二天就有宜家供货商来询盘，成交可能性很大。"一位中国展商跟法兰克福展览（上海）公司展会经理杨春激动地说。

杨春过年前就到了德国，出差快满三周了。虽然有些疲惫，但一有好消息，她都会分享，给大家鼓劲，朋友圈里是满满的正能量：

"第二天上午终于开张了，俄罗斯新客户，定了5个柜。"

"浙江一家做拖把的企业，自开展日起每天都有客户，昨天已经当场订合同，有一柜成交量。"

"一家生产便携式缝纫机和针线盒的企业，第一天没有客人，老板有点着急；到了第二天就接了一个很大的意向单，回去努力跟进即可达成；第三天、第四天都有实际签单；今天是撤展日，展商满怀信心地讲：希望今天能继续签单!"

开展三天时，单磊已经见了 30 多个客户，"以往交货期是 80 天，如果 2 月 17 日能复工，基本不会受什么影响"。

湖南一家给宜家、星巴克代工瓷器的老板更是雄心勃勃："工厂已经复工，给我两个月，订单没问题!"

至于崔翔勇的朋友圈，则贴出一张谈判本照片，三天时间，客户名片已经装满了整个本子。

"一切安好。危机中的机遇是实实在在的。中国的制造企业必能稳步渡过难关。"

（文章转自 2020 年 2 月侠客岛公众号，文章有删节，作者：李强）

思考并回答以下问题：

1. 本篇报道写在全球疫情爆发初期，记者通过采访德国法兰克福春季消费品展览会上的中国展商，体现了逆境中中国外贸企业抢抓订单的故事。请查阅相关资料了解德国法兰克福春季消费品展览会，这个展览会在国际上为何这么有影响力？

2. 文中提到了哪些中国制造的产品？这些公司为何一定要参加这个展览会？

3. 本文中提及中国外贸企业的产品出口给很多国际知名的大公司，如宜家、星巴克等，这种贸易方式叫 OEM，OEM 是英文 Original Equipment Manufacturer 的缩写，也称为定点生产，俗称代工生产，基本含义为品牌生产者不直接生产产品，而是利用自己掌握的关键的核心技术负责设计和开发新产品，控制销售渠道。讨论中国外贸企业在 OEM 模式下的优势与劣势。

4. "一家生产便携式缝纫机和针线盒的企业，第一天没有客人，老板有点着急；到了第二天就接了一个很大的意向单，回去努力跟进即可达成；第三天、第四天都有实际签单；今天是撤展日，展商满怀信心地讲：希望今天能继续签单!"通过这段描述，你对参加国际展会有什么感悟或者想法？你认为一场展览会上影响参展企业结识客户，获取订单的因素有哪些？

 **Reading 1**

Lead, Follow, or Get Out of the Way

You went to the show, wowed attendees with your exhibit, gathered hundreds of leads, and passed them on to sales. Your work here is done. Or is it? According to EXHIBITOR magazine's most recent Sales Lead Survey, at least 40 percent of leads generated on the trade show floor go unfulfilled. Essentially, four out of every 10 leads you collect in your exhibits might just as well go directly into the trash-along with nearly 40 percent of your program's value.

"Driving sales is why companies allocate big bucks to exhibit at events, " says Peter Gillett, CEO of Zuant, a lead-capture and-management firm. "If you don't follow up on leads properly, you are throwing money down the drain. Still, our anecdotal evidence suggests that up to 90 percent of exhibiting companies falter, at least to some degree, when it comes to following up." You might think it's not your responsibility to worry about whether or not the sales department does its job, and maybe you're right. But if your leads go unfulfilled, it's nearly impossible for them to translate into sales. And even if a sale does come through, if you're not tracking leads, your program is unlikely to get any credit, making the expense of exhibiting difficult to justify-especially if company management decides to curtail its marketing budget.

"Ultimately we do get evaluated, rightly or wrongly, based on how good of a job sales doesfollowing up on the trade show leads we provide," says exhibit-marketing consultant Bob Milam. "If the leads don't turn into sales, for whatever reason, it negatively impacts management's perception of our program's value."

So, to help you safeguard those all-important sales leads-while simultaneously underscoring the efficacy of face-to-face marketing-here are 10 ways you, as an exhibit manager, can improve your company's fulfillment rates. Follow these steps, and you'll not only lead the way to increased sales, but also be equipped to justify your company's investment and protect your trade show program in the process.

1. Proactively Plan to Follow Up

"To be clear, every company intends to follow up on sales leads, but the lack of an organized plan often derails those intentions," says Catherine Walker, director of exhibitor services for event-management firm Experient, a Maritz Global Events company. "Exhibitors carefully allocate the time needed to coordinate, travel to, and exhibit at a show, but they often don't schedule the time needed to follow up on leads. And if follow-up is not explicitly scheduled, 'real life' and other accumulated tasks often get in the way."

Mike Mraz, marketing strategist and educator for Skyline Exhibits, agrees. "Many marketing groups simply pass leads from the scanning machine on to the sales team, but there's no organized plan for following through," he says. "Sales and marketing need to get together on the issue of lead fulfillment because there's no way to pound that hammer unless there's accountability on the sales side as well."

Mraz suggests sitting down with your sales manager to determine exactly how and when sales associates will follow up with leads after the show. Will they call or email contacts? Will they fulfill requests for literature via email or direct mail or will the marketing department fulfill literature requests? And just how long after the show will this follow-up take place? This information is critical for two reasons. First, it allows you to be more specific with the promises you make to attendees, e. g. , "A sales associate will call you next week to answer your questions and email you the information you requested." Second, it creates a set of expectations you can use to gauge success or failure. For example, if the sales department agrees to follow up on leads via email within two weeks after the show, you can easily and objectively determine whether or not that expectation was met and hold delinquent or underperforming reps accountable. In Walker's words, "If your marketing team doesn't have a timeline or accountability for follow-up, you shouldn't be surprised when leads fall through the cracks."

2. Define Your Target Audience

A lead is little more than a business card if it doesn't provide information about the prospect, including his or her needs, buying influence, budget, and time frame. This information, usually obtained through a series of qualifying questions, is what helps you and your sales department learn about your prospects, customize communications, and prioritize follow-up. To help ensure that your sales department will actually pursue the leads you provide, the qualification criteria must meet their expectations.

"One of the main reasons leads don't get followed up on is that marketing and sales do not agree on the definition of a lead," Gillett says. "This must be addressed and corrected, or the company will continue to suffer lost revenue as a result."

According to Ruth Stevens, author of "Trade Show and Event Marketing: Plan, Promote, and Profit," the criteria used to define a qualified lead should be developed with direct input from the sales department. So involve sales reps in that discussion and allow them to tell you what is and is not important when qualifying and following up on a trade show lead. Once you know how your company defines a qualified lead, you'll know exactly what information you need to obtain from attendees in order to qualify them, and therefore what questions and fields to include on your lead forms or branching-logic lead-retrieval devices.

Mraz agrees, calling the sales team "an incredibly vital component" in the process of developing qualifying questions. He suggests working with sales to identify the five to seven pieces of information that are most important for salespeople to know, then crafting a series of qualifying questions to help staffers obtain that data from your booth visitors.

3. Appoint a Spy

Concerned that his exhibit-marketing strategy wasn't hitting its intended target, Milam hired three students to stand in the aisles surrounding his former company's exhibit at the Institute of Food Technologists show and conduct exit interviews with attendees. The students watched for prospects who interacted with members of his booth staff. Then, as those prospects left the exhibit, the students asked them questions about the company's marketing message. But what Milam discovered from his informal research had nothing to do with his overall strategy or the firm's key messages.

One of the last questions students asked was simply, "Did booth staffers make a promise to follow up with you after the show?" Based on the exit surveys, 850 attendees reported being promised some sort of follow-up action. But when Milam counted the number of leads with actual recorded promises, there were fewer than 150.

"What we discovered was that no one was writing anything down," Milam says. "Staffers weren't recording 80 percent of the promises they were making to attendees, so those promises were never being fulfilled. That was the real reason we weren't seeing as many show-related sales as management wanted-not because the show itself was bad or our strategy was flawed."

To alleviate the problem, Milam designated one of the company's administrative assistants as the Lead Sheriff. It was her duty to observe staffer/attendee interactions in the exhibit and make sure the leads got recorded, along with any promises made. The strategy increased the number of recorded promises from 150 to more than 700 in a single year.

4. Score Your Leads

Too often, unqualified leads are passed along to sales. In other words, instead of only giving themthe shiny needles, marketers mistakenly dump the entire haystack on sales reps. And according to Walker, that practice can negatively impact their perception of the value of trade shows in general, as well as the viability of any leads produced by face-to-face marketing.

Stevens advocates prioritizing leads by first working with sales to develop a scoring system that assigns a weight to each of your qualifying questions. Doing so means individual leads can be analyzed and given point totals that indicate how well they align with sales reps' predetermined criteria. For example, if their criteria are based primarily on budget and buying power, a lead from a decision-maker with a sizable budget would receive a higher score than a lead from someone without the authority or means to purchase your product. This scoring process helps prioritize the most valuable leads, along with those with the most potential to convert to a sale. Such prioritization means that even if sales reps don't follow up on every single lead you generate, at least they're more apt to follow up with the prospects most likely to make a purchase.

"In the modern digital world, this is absolutely vital," Gillett says. "Work with your sales department to hammer out an appropriate lead-scoring matrix, and that matrix will become your key to deciding which leads should be dropped, nurtured, or immediately issued to sales reps for prompt follow-up."

Additionally, multiple sources suggest scoring each lead and passing only the highest-scoring leads on to the sales department. The rationale behind this approach is simple: If you collect 100 leads, of which only 10 are qualified, you're asking sales to make 90 cold calls to people who don't meet the threshold for being a bona fide prospect. As a result, reps are less likely to take your sales leads seriously in the future, as their perception is that only a tiny fraction of exhibit-related leads are even worthy of follow-up in the first place.

5. Automate Emails

According to Ivan Lazarev, group head of event technology services at ITN International Inc., an Aventri company, there are four steps to effective lead management: capture, qualify, follow up, and distribute. "Most people skip follow-up because they confuse it with distribution," Lazarev says. "The follow-up step is simply making sure the attendee knows that you have registered his or her request. It's the 'Thank you for visiting our exhibit' email, and it should be done before the lead is distributed to sales." Thankfully, many modern lead-management systems allow you to automate that initial contact.

But what impact does this follow-up step actually have on lead fulfillment? Well, in a sense it instantly and automatically increases fulfillment rates to 100 percent, meaning that each prospect is receiving at least that autogenerated, post-show interaction with your company. But Lazarev says, at least anecdotally, that the follow-up step can also increase the likelihood of converting a trade show lead into a sale.

According to Lazarev, each of your follow-up emails should thank the attendee for visiting the exhibit, set a basic expectation (e. g. , "Someone will be in touch shortly to discuss our conversation at the XYZ show."), and include a link to a website where recipients can getadditional information about your company.

Gillett believes in a two-touch approach that occurs within 24 hours of a prospect's visit to your booth. "The standard practice is to wait for lead data to come through from a lead-retrieval company and then send out a thank-you email. But that data often arrives one or even two months after you've attended a show and is therefore virtually meaningless." Rather, Gillett recommends automating a generic thank-you email to be sent almost immediately after the staffer/attendee exchange and then following up a second time (within 24 hours) with more specific information that is customized to address the attendee's interests.

6. Determine Your Cost Per Lead

Without a clear understanding of the investment involved in generating trade show leads, it's hard for your sales reps to see them as more than names and basic contact information. To overcome this hurdle, Mraz advocates analyzing the total investment that each individual lead represents and then passing this information on to your salespeople.

"It's important for sales to know how much it costs to generate these leads," says Mraz, who claims that understanding the size of the investment inspires accountability. Determining your cost per lead is simple: If lead generation is your sole objective at a trade show, take the total cost of exhibiting at that event and divide that figure by the total number of leads you collected. That gives you the average cost to obtain each individual sales lead.

"If I spend ＄10,000 to exhibit at a show and I generate 100 sales leads, there is a ＄100 cost per lead. So, if I give you five leads to follow up on, that's ＄500 of our marketing investment, not just five pieces of paper," Mraz says. "If you don't follow up on those leads, it's the same as taking the company's digital camera and running it over with your car. You've destroyed an asset."

Assuming that you went to the show with multiple goals, assign each goal a percentage representative of its portion of your total objectives. For example, if your primary goal is lead generation, with secondary goals of boosting brand awareness and securing media exposure, lead generation may represent 60 percent of your total objectives, while the other 40 percent is split evenly between the other two goals. In that case, take 60 percent of the total cost of exhibiting and divide it by the number of leads generated.

7. Help Sales Avoid the Dreaded Cold Call

"Salespeople dislike following up on trade show leads that aren't vetted," says Kimberly Meyers, principal at marketing strategy and implementation consultancy Kimberly Meyers and Associates. "So, to help sales reps avoid the dreaded cold call and give them a good reason to pick up the phone, I've implemented a variety of show follow-up programs over the years." For example, when sales reps followed up on leads gathered at a print-industry show in Chicago, they invited prospects to a webinar on the power of variable-print technology.

According to Meyers, the tactic gave sales associates an initial talking point. In other words, rather than diving head first into a sales pitch, reps extended an invitation to the webinar before-they segued into follow-up sales discussions.

Another approach is to consider sending a direct mailer after the show but before the follow-up call from sales. Several years ago, one particularly savvy exhibitor sent each of the 200 qualified prospects it met with during a retail design, planning, and merchandising conference a direct mailer that contained a fresh orange. The mailer was an appropriate reminder of the company's exhibit, which featured 1,174 fresh oranges hanging from an overhead truss. The mailer also contained an accordion-fold card with pictures of the company's past design projects and a thank-you note with the firm's contact information. Sales reps followed up with each prospect within 24 hours of the mailers' expected delivery times.

The creative post-show promotion provided the perfect icebreaker for sales associates during their follow-up calls, as they were able to open conversations with, "Hey, did you get the orange that I sent you?"

8. Keep an Eye on Your Sales Leads

"If companies are serious about their trade show marketing investments, they have to hold the follow-up team accountable-almost at gunpoint," says Mraz, who encourages exhibit managers to keep tabs on which salespeople are notoriously lazy when it comes to following up on trade show leads. "If I'm giving you 10 leads, and you don't follow up on a single one, I'm not going to give you any more. I'm going to give them to people who are pursuing them and turning them into sales."

Other sources suggest distributing leads using a lead-or customer-relationship management (CRM) system, then tracking follow-up and creating reports that state how many qualified leads you generated, how many were handed off to the sales reps, how many leads they followed up on, and if possible which ones resulted in actual purchases.

Milam has developed a similar reporting procedure. The same administrative assistant that served as his Lead Sheriff was later tasked with tracking trade show leads in the CRM database and creating a monthly report. "I think most exhibitors do at least a decent job at trade shows; they are producing business opportunities for sales to act on. But people continue to question the value of exhibiting. The criticism falls on us, but the responsibility for success lies elsewhere," Milam says. "If the value of your program is called into question, and you've actually been tracking your leads, then you can legitimately pass some of the blame to where it belongs. But you need to have the numbers that prove others are not doing their jobs and following up on the sales leads your program is producing."

9. Nurture Seemingly Dead Sales Leads

What happens to a lead that doesn't turn into a sale? According to Stevens, many are discarded altogether. But even these so-called dead leads still provide value in the form of potential future sales.

"Studies have shown that 45 percent of business inquiries eventually result in a sale," Stevenssays, citing a report by the Marketing Management Journal as the basis of this rule of thumb. "They may not be ready to buy right now, or maybe they don't currently have the budget to do so, but eventually they're going to make a purchase. You need to stay in touch with them so that when they are ready to buy, they turn to your company instead of purchasing from one of your competitors."

Referred to as lead nurturing, lead incubation, lead recycling, and lead development, this i-dea represents a deliberate marketing strategy of repeated communications (via phone, email, and/or direct mail) designed to keep your company top of mind. And according to Stevens, that task falls on marketing's shoulders.

"Marketing's job is to deliver qualified leads to sales. If a lead is not qualified yet, that nurturing process belongs in the marketing department," Stevens says. "Marketing should assess the leads after the show, give the ones that are qualified to sales, and begin nurturing and developing those that are not yet qualified. Once they become qualified leads, perhaps by requesting a quote, that's the time to pass them on to sales."

While there are no hard-and-fast rules for how or when these communications should take place, Stevens suggests varying their regularity to avoid making them look like autogenerated e-blasts. "Email them a case study about how an existing client has used your product or service, call and invite them to a VIP event, or send them an announcement about a new product you're debuting," Stevens says. "Don't just send a 'Hi, how are you?' email. Send them useful information to help move them along the decision-making process."

10. Commit to Making Ongoing Improvements

"Improvement in your organization's lead qualification and fulfillment is an ongoing process," says Gary Survis, venture partner at Insight Venture Partners. "Every sales organization wants its sales funnel to have a large prospect pool, especially ones that experience longer sales cycles. So, to succeed, you need to put a stake in the sand, set a goal, and make sure your sales department is taking advantage of the leads that your program is bringing back to the table."

Start by establishing and circulating follow-up goals. For example, if you suspect sales is following up on a mere 40 percent of the leads you collect, set a goal of 50-percent follow-up for your next show and implement the tips and techniques that make the most sense for you and your organization. If you reach your goal, pat yourself-and your sales team-on the back. Then raise the bar bit by bit at each of your events to keep the ball rolling on the path toward 100-percent fulfillment.

If you don't meet your initial goal, seek internal allies such as sales or marketing directors who might be willing and able to help you crack the whip. Or implement a few different tactics. But do something-anything-other than sitting idly by as your sales reps doom your program, and potentially your career, to an ill-fated end.

After all, you basically have three choices: You can resign yourself to the status quo, you can cross your fingers and pray things get better on their own, or you can make an attempt to improve your company's follow-through. Even if your solution fails to completely correct the problem, you'll likely see a slight improvement that will get your program moving in the right direction. As the saying goes, you need to lead, follow, or get out of the way. So, lead the way toward consistent lead follow-up, and increased sales will hopefully follow.

(This article is edited from Internet information)

NEW WORDS AND PHRASES

attendee　n. 参观展览会的人(不包括参展商)

sales lead　潜在客户, 销售线索

exhibit　n. 展品

exhibitor　n. 参展商

anecdoted evidence　轶事证据, 坊间传闻的证据

derail　v. 破坏, 阻碍

schedule　vt. 安排, 列入

accountability　n. 责任心

fulfill request　满足要求

literature　n. 宣传资料

expectation　n. 期望值

gauge　vt. 预测, 预估

delinquent　adj. 失职的

underperforming　adj. 表现不佳的

fall through　落空, 失败

buying influence　购买影响力

time frame　时间段

prioritize　v. 按优先顺序列出

prospect　n. 潜在客户

address　vt. 解决, 处理

lost revenue　收益损失

incredibly　adv. 难以置信地, 非常地

booth visitor　摊位参观者

exit survey　离场调查

alleviate　vt. 减轻, 缓和

viability　n. 可行性

weight　n. 权重

align with　与……一致

matrix　n. 模型, 矩阵

rationale　n. 基本原理

bona fide　忠实的, 真诚的

skip　v. 不做(应做的事)

post-show interaction　展后互动交流

generic　adj. 通用的

hurdle　n. 障碍, 困难

dreaded　adj. 可怕的, 令人畏惧的

old call　陌生电话, (向潜在客户打的) 没有预约的电话

webinar　n. 网络研讨会

sales pitch　推销商品的言辞

segue　vi. 接入,转入

savvy　adj. 有经验的, 聪慧的

icebreaker　n. 活跃气氛的话,消除隔阂的行动

keep tabs on　密切注视,监视

legitimately　adv. 合理地,正当地

discard　v. 丢弃,扔掉

the rule of thumb　经验法则

hard-and-fast　固定的

regularity　n. 规律性

announcement　n. 商品简讯

debut　v. 首次推出

sales funnel　销售漏斗

pat on the back　表扬,激励

get out of way　闪开

resign　v. (使)顺从

status quo　现状

NOTES

1. (sales) lead，潜在客户或销售线索。销售线索一般是指通过会展、电话咨询、消费者访谈等多种方式获得销售的初级线索，销售人员再持续跟进和推动线索的继续延伸，到达成熟阶段后销售线索转换为销售机会。销售人员将销售机会进行漏斗式管理和推进，经过谈判、产品和技术沟通，最终和客户成交，订立合同。

2. sales funnel，销售漏斗。销售漏斗是跟踪客户从潜在客户转化为忠诚客户的过程的一种方法，它通常表示为一个销售对象在成为客户过程中所经历的一系列阶段。典型的销售漏斗从潜在客户或线索生成开始，然后通过与销售团队的一些互动来跟进潜在客户，最终促成客户转化。销售漏斗的最后阶段侧重客户保留和忠诚度。

3. a/the rule of thumb，经验法则。一般来说这些法则是凭经验所得来的方法，并不要求完全精确，这个表达一般以单数形式出现，a rule of thumb or the rule of thumb。

4. customer-relationship management（CRM）system，客户关系管理系统。客户关系管理系统是利用信息科学技术，实现市场营销、销售、服务等活动自动化，是企业能更高效地为客户提供满意、周到的服务，以提高客户满意度、忠诚度为目的的一种管理经营方式。客户关系管理既是一种理念，又是一种软件技术。以客户为中心的管理理念是 CRM 实施的基础。

EXERCISES

Ⅰ. Answer the following questions.

1. What percentage of leads generated on the trade show floor go unfulfilled according to EXHIBITOR magazine's most recent Sales Lead Survey?

2. How many ways does the author recommend to improve your company's fulfillment rate?

3. What is the duty of Lead Sheriff?

4. According to Ivan Lazarev, what is the function of follow-up step?

5. Please list two approaches to help sales reps avoid the cold call?

Ⅱ. Please translate the following English sentences into Chinese.

1. If you don't follow up on leads properly, you are throwing money down the drain.

2. To be clear, every company intends to follow up on sales leads, but the lack of an organized plan often derails those intentions.

3. Additionally, multiple sources suggest scoring each lead and passing only the highest-scoring leads on to the sales department.

4. To help sales reps avoid the dreaded cold call and give them a good reason to pick up the phone, I've implemented a variety of show follow-up programs over the years.

5. Marketing should assess the leads after the show, give the ones that are qualified to sales, and begin nurturing and developing those that are not yet qualified.

6. Driving sales is why companies allocate big bucks to exhibit at events.

7. If your marketing team doesn't have a timeline or accountability for follow-up, you shouldn't be surprised when leads fall through the cracks.

8. A lead is little more than a business card if it doesn't provide information about the prospect, including his or her needs, buying influence, budget, and time frame.

9. This must be addressed and corrected, or the company will continue to suffer lost revenue as a result.

10. Work with your sales department to hammer out an appropriate lead-scoring matrix, and that matrix will become your key to deciding which leads should be dropped, nurtured, or immediately issued to sales reps for prompt follow-up.

Ⅲ. There are 10 sentences in this section. Beneath each sentence there are four words or phrases marked A, B, C and D. Choose one word or phrase that best completes the sentence.

1. Which of the following is the best translation of the phrase "a rule of thumb"? (　　)

A. 拇指条例　　　　B. 经验法则　　　　C. 行事规则　　　　D. 拇指准则

2. In the time schedule of exhibitors, which of the following is easy to be left out? (　　)

A. the time needed for travel　　　　B. the time needed to coordinate

C. the time needed to exhibit at a show　　D. the time needed for follow-up of sales leads

3. Which of the following gives the closest explanation of "cold call"? (　　)

A. a call to potential customers by sales reps without any prior contact.

B. cool call

C. a call given to customers after the show

D. a call made by exhibitors to attendees

4. Which of the following can help us to decide which leads should be dropped, nurtured, or issued to sales reps for follow-up? (　　)

A. qualifying questions　　　　B. sales leads survey

C. expectations　　　　D. lead-scoring matrix

5. Which of the following method can help sales avoid cold calls? (   )

　　A. direct mailer　　　　B. sales letter　　　　C. follow－up　　　　D. advertisement

6. Which of the following is the best translation of the phrase "sales funnel"? (   )

　　A. 销售漏斗　　　　B. 销售烟囱　　　　C. 销售斗状物　　　　D. 销售通道

7. According to Milan, (   ) percent of the promises they were making to attendees, so those promises were never being fulfilled.

　　A. 90　　　　　　　　B. 70　　　　　　　　C. 80　　　　　　　　D. 50

8. In order to work out an organized plan for following through, marketing team needs to get together with (   ) on the issue of lead fulfillment.

　　A. exhibitor　　　　　　　　　　　　B. sales team

　　C. management personnel　　　　　　D. HR manager

9. There are four steps to effective lead management: capture, qualify, follow up and distribute. Among the four steps, which one is likely skipped by most people? (   )

　　A. capture　　　　B. qualify　　　　C. follow up　　　　D. distribute

10. Which of the following is the best translation of the phrase "CRM system"? (   )

　　A. 客户关系管理系统　　　　　　　　B. 客户关系管理制度

　　C. 客户关系管理体系　　　　　　　　D. 反雷达导弹

Ⅳ. Writing

Please write a plan to increase the fulfillment rate of sales lead.

📖 Reading 2

How to Quantify Your Target Audience

　　After my company missed its on－floor sales goals at a show for two years running, I decided to learn more about the event's audience. What I discovered changed my whole mindset, as I realized we were truly targeting just 1 percent of the show's 22,000 attendees. This new awareness resulted in my team redefining our goals and objectives, asking for a cut in our show budget, and more than doubling our return on investment.

　　Numbers don't lie-but they can obscure the truth. And taking the gross number of attendees at face value is a misstep too many exhibitors make. A few years ago, a medical－supply company that specializes in eye－care equipment hired me to be the event manager for its trade shows. As anyone in a new position should, I took some time to figure out what the company was doing at its shows and why. One immediate point of concern was that for two years in a row my new employer had failed to meet its on－floor sales goals and get a satisfactory return on investment at an important optometry show that regularly drew more than 20,000 attendees. Being tasked with overseeing this program, I felt I needed to pinpoint the problem and come up with a course correction.

　　After crunching the numbers and poring over the attendee data, I found that we were casting too wide a net for too few fish, i. e. , we weren't correctly quantifying our target audience and were overspending at a show without enough potential revenue to hit our targets. As a result, I suggested to my leadership that we rethink our expectations, reduce our exhibit's footprint, and reallocate the savings to an event where the money would be used more effectively. Here, I'll explain how I went about quantifying our target audience and, in turn, set goals and objectives that reflected reality.

➤ Request Information From Show Management

I had an inkling that our challenges were rooted in overestimating the number of attendees that we were targeting. So, I reached out to our association, asked if it could help me understand who attended the event, and received the audited attendee analysis seen here. About 7,200 of the 22,000 showgoers were exhibitor staff and media, which left about 15,000 nonexhibiting attendees. That's still a massive number and nowhere near our true target audience, but I was making headway by narrowing the funnel by roughly 30 percent.

➤ Identify Which Attendees Align With Your Business

Now I knew I was fishing in a pond stocked with 15,000 nonexhibiting attendees, but I still didn't know much about them. So, I reached back out to my contact to see if I could get a clearer picture of just who was walking the show floor. The association sent a list of attendees from the previous year broken down by the type of business they were in. The data showed that the vast majority of visitors were retailers or involved in other practices that disqualified them from being in our target audience. Our key buyers at the show are ophthalmologists, of which there are about 1,000 or less than 7 percent of the total attendance.

➤ Set Realistic Expectations

Now that I had a grasp on how many actual prospects would be at the event, I began setting more realistic goals for just how many we could expect to reach. Based on my experience, I figured that half of the 1,000 ophthalmologists wouldn't be in the market to purchase goods within the next 30 days or simply wouldn't be interested in our product, leaving me with 500 candidates. Then I cut that number in half again because we were one of six exhibitors competing for the same audience, and it was safe to assume that a good chunk of those prospects either had relationships with our competitors or would be visiting-and possibly inking deals-with them. By my rough math, the 22,000 "prospects" the association told me I had were now winnowed down to a mere 250, i. e. , about 1 percent of the total show attendance.

➤ Rethink Your Investment and Objectives

Historically, our presence at this show comprised a 20−by−30−foot booth with a budget of almost $150,000. Staffers took orders in the exhibit, and our on−site sales goal was in the neighborhood of $1 million. Since our products' average price tag is in the low six figures, we had to close nearly 10 sales at the show to reach that target-which, as I said, hadn't happened in two years. And seeing as how we were actually fishing in a pond of 250 instead of an assumed ocean of 22,000, I wasn't too surprised the company was missing its million−dollar mark. So I went to my stakeholders, presented my findings and figures, and recommended downsizing to a 10−by−20−foot space, sending three staffers instead of the usual six to eight, and setting the more realistic objectives of netting 12 quality leads and landing one deal on the show floor.

➤ Find Your Fit

The process of quantifying your true target audience takes time and effort, so it may not be feasible for you to complete this task for every event on your calendar (exhibit managers just don't have that kind of time with all the other tasks on our plate), so I suggest starting with one or two of your key or troublesome shows. I bet you'll find the payoff and the potential savings are well worth the effort. My company likely would not have implemented these changes if I hadn't homed in on how much-or little-of our target audience was actually present at the show. But reevaluating our approach led to us trimming our investment in this show by $100,000, more than doubling our ROI, and, much to my delight, achieving our sales target. Yes, the goal was more modest, but when you factor in the reduced investment, we still pleased our stakeholders. Even better, we reallocated our six-figure savings to other events where there were bigger fish to reel in.

(This article is edited from Internet information)

 NEW WORDS AND PHRASES

miss on-floor sale goal　没有达到场内销售目标

mindset　n. 思维定式

return　n. 回报

obscure　n. 遮掩, 掩盖

the gross number　总量

take at face value　根据外表判断, 根据表面现象判断

oversee　vt. 监督

pinpoint　vt. 查明, 找到

crunch　v. 处理(数字)

pore over　集中精神阅读, 仔细阅读

overspend　v. 花费过多, 超支

overestimate　v. 过高估计

a good chunk of　一大块儿, 大量的

winnow down　筛选, 挑选

comprise　vt. 包含, 包括

on-site sale goal　现场销售目标

stakeholder　n. 利益相关者

net　v. 设法获得

land　v. (轻而易举地或意外地) 获得

feasible　adj. 可行的, 办得到的

payoff　n. 报酬, 回报

potential saving　潜在的开支节省

home in on　跟踪

ROI(Return On Investment)　投资回报率

trim　v. 削减, 减少

factor in　把……考虑进来

reel in　收卷掉线, 钓起来

Reading 3

Made to Measure: A Conversation with Joe Federbush

How has the pandemic changed the way face-to-face marketers should approach metrics and tracking their programs' performance? Measurement maestro Joe Federbush warns that some familiar strategies are less relevant on post-pandemic trade show floors, but that the evolving landscape is teeming with new and exciting opportunities.

Most exhibit-marketing managers have probably heard some variation of the adage, "You can't improve what you don't measure." But merely knowing you should have a measurement plan is a far cry from implementing one. That's because there are hundreds of potential metrics and measurement tools to choose from, creating the threat of "data paralysis". Thankfully, Joe Federbush, president and chief strategist of Evolio Marketing Inc. , has some concrete insights to help navigate the measurement minefield.

For more than 20 years, Federbush has helped companies develop and implement effective performance-tracking plans, and he recognizes that the rapidly evolving measurement landscape can overwhelm veterans and newbies alike. From KPIs and OKRs to ROI and ROX, the topic is rife with acronyms that can leave an exhibit manager's head spinning. Here, EXHIBITOR asks Federbush to explain how the pandemic fundamentally altered which metrics exhibitors should be homing in on, what specific strategies they should be implementing, and why we should be excited about a new generation of analytic tools.

Exhibitor Magazine: In-person shows are coming back, and early reports indicate a general decline in attendance but an increase in lead quality. How should exhibitors factor this into their measurement efforts?

Joe Federbush: It sounds cliché, but it's probably truer than ever: Exhibitors need to focus on quality over quantity. Since the start of the year, trade show attendance has been generally down 30 to 40 percent. To make matters tougher, a recent survey we conducted revealed that 92 percent of show organizers report attendees are waiting until the last minute to commit to an event, meaning exhibitors don't even know who will be at a show. This means exhibitors need to recalibrate their measurement strategy. In the past, program managers have been setting objectives based on total attendance numbers, which just isn't going to cut it today. They need to pay much closer attention to attendee demographics and map out exactly who their true target audience is. It's a mind shift that really should have taken place years ago.

EM: How has the pandemic affected our industry's collective attitude toward measurement and the types of metrics that are prioritized, such as return on investment (ROI)?

JF: ROI is going to remain a critical goal, and in some ways will be even more important moving forward since budgets are still going to be tight for a while. Executives want to see that a show is worth the time and money put into it. That said, I expect exhibit programs to shift their focus to the importance of return on experience (ROX). Experiential marketing was already important before COVID, and I expect that to continue.

Creating a positive ROX begins with working through show organizers to determine what products and technologies attendees expect to see and then developing an experience that aligns with those desires. Exhibit managers will want to focus on exit and post-show surveys to track and assess ROX. Develop one or two statements on a five-point scale from "strongly agree" to "strongly disagree" about whether the booth's information, offerings, and displays met or exceeded visitor expectations.

Another key ROX metric to track through surveys is staffer performance. Exhibit managers should measure the quality of staffers' engagement, interactions, and knowledge to ensure their teams are helping drive stellar experiences. Once you gather the data and assess it, work to improve from one event to the next.

EM: The other topic everyone's talking about is hybrid. What advice would you give to an exhibitor who is worried about tracking their company's in-person and online performance?

JF: Hybrid is an interesting topic. In March, 53 percent of show organizers said they are going to include a digital component, so expect about half of events to be hybrid in some capacity. However, unless you're investing a lot of time and money in the virtual component, don't expect it to generate a lot of leads or traffic.

Trying to track in-person and online components at the same time can be daunting, but it doesn't have to be. My advice is to be conservative in terms of both goals and measurement. Start by setting five to eight goals for the in-person component and do the same for online, and then create a feasible measurement strategy to track your success. And don't worry if it's not perfect right out of the gate. As the saying goes, "Don't let ' perfect' become the enemy of ' good. ' "

EM: By all accounts, virtual events aren't disappearing any time soon. Everyone got a crash course in 2020, and attitudes and "hot takes" on virtual-event metrics have evolved rapidly. Where do things now stand in terms of the key metrics to focus on?

JF: The upside of virtual events is that nearly everything is trackable, and the downside of virtual events is that nearly everything is trackable. It's easy to become overwhelmed by the vast amount of data available, so the key is to focus on what's relevant to your objectives. From what we've learned so far, digital events are not stellar at creating networking opportunities or generating new leads. But they do a pretty good job at delivering other benefits: building brand awareness, promoting thought leadership, and delivering valuable information. Focus on tracking the success of those.

If you're still feeling stuck, consider four key metrics. One, collect the total number of visits and the total number of unique visitors. Take a look at who came, how long they stayed, and who returned multiple times. This will give you a good idea of who's really interested in your offerings and who just popped in for a quick look. Plus, if you have a goal about the number of target attendees you hope to engage, you'll have a clear figure with which to measure the success of the event.

The second metric focuses on the value of the experience ROX again which can be measured via polls and surveys. For example, ask attendees to rate the usefulness of the presented information and/or their agreement with key messages. Three, measure brand impact. This can be done using a pre-and post-event questionnaire that asks about interest level or impressions of your company and will let you know whether your online endeavor boosted brand awareness. Along the same lines, consider using the Net Promoter Score (NPS), which is very relevant to tracking brand sentiment at virtual events. Finally, measure business impact. After leaving your event or experience, are attendees motivated to schedule a meeting, download a white paper, visit your website, or do further research about your products?

EM: What are the most cost-effective measurement tools and methods that deliver insightful feedback?

JF: My favorite measurement tool is literally free: observation. Some exhibit managers never take the opportunity to step outside their booths and just watch. Start with the basics: Is your brand clearly discernible from a distance and up close? Are your products readily identifiable from the aisle? Are there several visitors that look like they want to talk with someone but all the staffers are engaged since you didn't bring enough? Simple observation is a powerful tool that doesn't cost a nickel but is often overlooked.

If you have a little money to spend, surveys give you the best bang for your buck. Develop a measurement plan and standardize reporting templates that can be replicated across events. After the up-front expenses and resources, the cost of research drops significantly over time. Plus, if you have consistency across shows, you can start to develop your own benchmarks and compare results across events to start making better portfolio, investment, and staffing decisions.

EM: What do you see as the biggest challenges for exhibit managers when it comes to collecting, analyzing, reporting, and acting on data?

JF: I really don't think it's a lack of knowledge-it's a lack of bandwidth and resources. But that doesn't mean it cannot be done. Take baby steps and start with your tier-one shows where you have more money and time invested. If you cannot do that effectively internally, look at outsourcing to an expert that can tailor a measurement program and guide you through the process to ensure you're getting the feedback you need.

Another challenge is that exhibitors are pretty bad at cutting out shows that no longer produce strong results and hit what I call the 80/20 rule: 80 percent of their time is spent working on tier-three events that only account for 20 percent of their budget, and they only spend 20 percent of their time focusing on the tier-one shows that gobble up 80 percent. A strong measurement game can really help a team identify where they should be spending their efforts and what shows need to get dropped completely.

EM: How do you see measurement evolving in the coming years?

JF: One of the biggest leaps we're seeing from a market research and measurement perspective is the growth of artificial intelligence (AI). Research companies are coming up with ways to use AI to do dive-deep analysis, especially when it comes to finding patterns in things like open-ended feedback. In the coming years, AI is going to become more and more affordable, which will be a huge benefit to those collecting data.

The other area impacting marketers is the growth of data visualization tools and software. It's an important sector to watch because it makes everyone look like a data genius and makes it effortless to communicate your data story to stakeholders. It's an exciting time to be measuring and analyzing data, and it's never been more important for companies to stay on top of their measurement game.

(This article is edited from Internet information)

 NEW WORDS AND PHRASES

metrics   n. 度量标准

maestro   n. 大师

measurement plan   检测方案

a far cry from   相差甚远, 悬殊

concrete insight   具体的见解

performance-tracking plan   业绩追踪方案, 业绩评价方案

rife with   充满着, 充斥着

newby   n. 新手

recalibrate   v. 重新调整, 重新校准

attendee demographic   展会参观者统计资料

mind shift   观念变革

executive   n. 高管, 主管

return on experience (ROX)   体验回报

experiential marketing   体验营销

staffers' engagement   员工参与度

traffic   n. 信息流量

daunting   adj. 令人生畏的, 使人畏缩的

by all accounts   据说, 据大家说

hot take   脱口而出的评论

upside   n. 优势, 好的一面

downside   n. 劣势, 不好的一面

networking opportunity   交流机会, 建立关系网的机会

pop in   偶然来访, 突然出现

poll   n. 民意调查, 投票

Net Promoter Score (NPS)   净推荐值

brand awareness   品牌知名度

brand sentiment   品牌情绪

discernible   adj. 看得清的, 可识别的

overlook   vt. 忽略

up-front   adj. 前期的

template   n. 模板, 样板

portfolio   n. 投资组合

outsource   v. 外包

ENGLISH FOR WORKPLACE COMMUNICATION

Sample Dialogue: *A conversation at a Trade Show.*

Marie and Yuca are attending a trade show in San Diego. Maria is from France and Yuca is from Japan. They've met before at last year's trade show. Let's listen to their conversation.

Maria: Hi Yuca, it's good to see you again. Are you enjoying the trade show?

Yuca: Hello Marie. Nice to see you too. Yes, I've seen a lot of interesting new products and gotten some good ideas. How about you?

Maria: Same here, I had lunch at a Mexican fast casual place yesterday. Maybe you'd want to have lunch there later?

Yuca: Sure, how was the food?

Maria: The tacos were really good. One thing surprised me though. I ordered at the counter and paid at one of those payment tablets. One screen asked me how much of a tip I wanted to leave 10%, 15% or 25%. I don't remember seeing that before.

Yuca: It was taken out, right? That seems odd. Even here in a sit-down restaurant you have to leave a tip of course and 15% or 25% seems like the expected range. Though it seems pretty steep to me in Japan we don't leave tips at all. But it seems pushy to suggest such a big tip at a fast-food place. I guess customs change.

Marie: They certainly do. There really are a lot of customs that are different here.

Yuca: I know like calling everyone by their first names. Yesterday I met the CEO of ABC company Bob McDonald. I called him Mr. McDonald but he said everybody calls me Bob in Japan. I just wouldn't feel comfortable calling a CEO by his first name or anyone much older than me for that matter.

Marie: Yeah, I remember the first time I was in the US. Meeting new people was awkward. I wasn't sure if I should shake hands or kiss them on the cheek or just give a little wave.

Marie: I know what you mean. Sometimes I've started to bow, but the other person was putting out their hand for a handshake. And it used to bother me when I gave my business card to someone and they just put it in their pocket. That seems so disrespectful. But since then, I've realized they don't mean to be rude. It's just the way people do things here.

Yuca: And saying goodbye is confusing to people say we'll have to get together sometime or let's do lunch. But they don't really mean it, they're just being friendly.

Marie: So, do you really want to have lunch today?

Yuca: Sure, I'll meet you at noon at the west entrance to the convention center.

Marie: Sounds good, I'll see you then.

本章导读

　　展后跟踪是参展的第三个阶段，也是能否促成最终成交的关键一环。了解跟踪展会的重要性，特别是如何与潜在客户进行沟通和交流。展后跟踪还包括准备报价单和样品以及订单生产和发货等，尽管这些属于国际贸易的范畴，但也是展后跟踪的主要内容。

　　展后跟踪要有计划、有目的，除在公司对展会进行复盘外，更重要的是针对不同的客户进行有针对性的跟踪。邮件、社媒、电话等都是跟踪客户的有效方法，但需要掌握一定的技巧。通过本章的学习，掌握跟踪一场展会需要做好的工作，特别是细节要做到位。

知识精讲

展会知识

外贸人，都应该学学她

　　李雪，国外的客户知道她叫 Jessica K K，是全全全科技河北有限公司的销售经理，她创造了一个客户一个订单 300 个货柜的历史，外贸人都应该学习她，下面分享她在外贸路上成长的故事。

　　2013 年李雪参加了在法国巴黎举办的国际建材展览会（Batimat），在这个展览会上结识了一名阿尔及利亚的客户（法国建材展有超过三分之一的买家来自北非等国家和地区）。阿尔及利亚的官方语言是法语，很多时候即使他们会说英语都不愿意讲。李雪根据经验判断，这个客户对护栏类的产品非常精通，而且之前在中国买过不少货。在和客户的交流中，她能感觉到这位客户关注点在于价格和质量，当时客户并没有太重视李雪，在李雪给他报价的过程中，客户一直是很高傲的态度，强调价格太高了。

　　展览会结束后，李雪开始跟踪这个客户，但客户回复邮件只用法语，从不用英语。如果换作他人或许这个客户就黄了，但性格执着的李雪为了能和这个客户更好地沟通，开始自学法语，勤奋和执着让李雪的法语进步很快，经过近一年的学习，她已经能用法语熟练地给客户回复邮件了。就这样，李雪通过邮件和客户保持着不间断的联系。

　　2014 年 10 月的广交会。客户来到中国，应邀和李雪见面并考察工厂，在此期间，客户也终于敞开心扉提出了他心中的种种疑虑，比如他所接触的中国工厂报价都很低，而承诺的质量却达不到要求。李雪认真解释并对比了很多产品的不同，带客户考察了不同的工厂，让他全面了解了护栏这一产品在中国的现状。

广交会结束后，客户给李雪发来了厚厚一本产品报价单，大约有 50 页，而这些文件全部都是法语的。李雪用了两周的时间翻译了全部的产品规格和具体要求，这次翻译过程也让她学会了如何用法语表述几乎所有的护栏类相关产品。

两年很快过去了，2015 年的 11 月，李雪再一次参加法国建材展（法国建材展每两年举办一届，单数年的 11 月上旬举办），并再次与客户碰面，而这次带给李雪的是真正的一个大订单。两年的磨合、交流和沟通，让李雪收获了 300 个货柜、一年交货期的大订单。

根据李雪的介绍，这个客户的订单还在继续，每年都不少。这不，此时此刻李雪正在阿尔及利亚参加建材展（每年 3 月），不知道她这次又能收获多少订单呢？

（本文转自 2019 年爱德会展公众号）

思考并回答以下问题：

1. 查阅相关资料了解法国巴黎国际建材展览会的总体情况，包括主办方、展出面积、主要展出商品、主要采购商来源地等。

2. 本文记录了李雪在一场展览会后如何跟踪一位来自阿尔及利亚客户的故事，读完后，你认为李雪的哪些做法让她赢得了 300 个货柜的订单？

3. 外贸企业参加国际性展会的连续性很重要，你是如何理解参展连续性的？

4. 讨论展后跟踪的重要性。

（注：可以关注爱德会展公众号，找到崔老师随笔系列文章，闲暇时间读一读。）

 **Reading 1**

Best Practices: How to Convert Leads with Trade Show Follow-up

According to Salesforce, the world's most popular customer relationship management software, a shocking 80 percent of exhibitors don't follow up with leads after a trade show. In one of their recent studies, if a salesperson reached out at all, it often took 50 days. It's likely that after that much time has passed, the lead has forgotten your name, if not lost interest in your brand altogether.

If your salespeople aren't communicating with leads after the show, not only are they losing opportunities and potential revenue, but the time and money the company invested in being at the trade show is wasted. For you as the trade show manager, it is important your sales colleagues understand you are responsible for the overall success of your program, not just the trade show booth design.

Whether you're a trade show veteran or you're new to it, there are always ways to improve your trade show tactics. After all, there are tons of opportunities that are present at trade shows:

49% of trade show attendees intend to buy one or more of the products/services exhibited.

82% of these attendees have the authority to buy on behalf of their organization or business.

A trade show gives you direct access to these key decision makers. But it's not just about making contact or showcasing your product, it's about how you follow up with your leads post-show. That's how you close the sale.

Trade Show Pre-planning

There's much to be said about post-show actions and trade show follow-ups. But before you can think about post-event emails and lead follow-ups, you need to make sure that you're on track months before your trade show arrives.

There are definitive marketing techniques for trade shows that one should follow leading up to the event. We also suggest creating a timeline for yourself so that you can stay on schedule and don't miss any important details of the planning process.

Make sure you start planning months ahead, don't leave yourself with little to no time to plan what your on-site call to action will be, or how you plan to steal the show with an amazingly designed exhibit.

Ensure that you have clear goals and understand why you're participating in that particular trade show. Plan what it is that you'd like to tell potential customers when they stop at your booth. Simple informative brochures and catalogs in this day and age won't work.

Get the word out. Make sure you start telling people way ahead of time that you'll be exhibiting. Harness the power of your social media audience and do a build-up campaign to the trade show. Perhaps even run a special offer for those who follow your social media and visit your booth, not only will this encourage them to visit your display, but you could use it was a way to measure the impact of your social media campaign.

Capture Your Leads

You can't launch a follow-up campaign without any leads. Before you can action a post-show strategy and follow up, you'll need to gather some information from your booth visitors.

➢ Design Your Exhibit Strategically

There's no point in placing a brochure rack right at the entrance to your exhibition stand. Hordes of people will simply walk past, grab a brochure and keep walking. You will want to make sure that you spend enough time thinking about the exhibition stand design and build before you set anything up first. You will want to make sure that you interest people enough so that they come to your stand.

You don't want to give show attendees the opportunity to pass you up. When designing your booth, think about doing it in such a way that people are invited to come inside, to interact with you and your brand. This way they're more likely to interact, ask questions and leave their details for follow up communication.

➢ Perform a Game or Do a Giveaway

Ever seen a prize wheel ready to spin? Human nature has you wanting to walk over and spin it. Entertain your booth visitors by "gamifying" your exhibit and making it more interactive for people who choose to stop in.

Another great idea is to advertise a simple giveaway that entices trade show visitors to walk in, grab a little treat and perhaps leave their business card in the box.

➢ Your End Goal

If your end goal is an email follow-up campaign, then obviously your goal during the trade show is to collect email addresses and other contact information. Make this a main priority and figureout the best way for you to collect those email addresses. Your next step will be to then add them to your mailing list as soon as possible, so you can keep them up to date with your business news. Although, before you get too enthusiastic about this step, you may want to learn how email verification works so you can make sure that your emails are actually reaching "real" people and not invalid addresses. This could help you in the long run.

Host a Recap Session

To ensure everyone is well informed after the event, hold a meeting that includes your marketing team and sales leaders. You should cover key take aways from a marketing perspective, and give sales reps the opportunity to do the same.

This is also a good time to train your reps for how to speak to leads. This shouldn't feel like a patronizing session, but rather a time to use the same message about your brand across the company.

In the same light of continuity, walk through your CRM or database so all reps are using and understanding it the same way. Should they take notes in it? If so, where? How often? How do they change the status of a lead from cold to hot? All of these points should be covered so everyone is on the same page.

Be Social

Connect with your leads on social media. Like, comment, follow and engage with their pages and posts. Do this immediately following the trade show and you increase your brand's chance at being remembered.

Trade Show Follow-up Emails

"A man who does not plan long ahead will find trouble at his door." Confucius, Chinese philosopher said. You're probably thinking, "OK, I've collected tons of email addresses…what next?" Here's a tip: immediate action is key.

Don't wait until after the trade show to decide what your follow-up email message is going to be. You'll want to plan this strategy well ahead of time. You should have the email template designed and ready to launch just a few days after the show. Even a day or two after the show will set you miles ahead of your competition who may take weeks to send out their first line of post-show communication.

Your plan should be to make the first contact. Odds are that there were a few businesses at the trade show that do exactly what you do, or sell exactly what you sell. Being the first to make contact already puts you a step ahead of your competitors.

Post-event email is a great way to make that first contact and potentially turn your lead into a customer. But make sure that you don't use the same email for different trade shows. Consider putting the name of the show in the subject line, and even personalize your email campaign.

Email one: This is the initial message you send within the first two days after meeting them at the event. Include an image of you and other team members at the booth to help them recall who you are. Thank them for visiting the booth or speaking with you at the trade show. Consider giving them a special offer for visiting the booth.

Email two: Schedule this email for one to three days after the first email. Include a clear CTA, such as inviting them to take a survey about their experience at your booth, watching a video about the company and product or service or signing up for a virtual workshop or event.

Email three: This email is scheduled three to four days after the second email. Include another CTA to get them to visit your website or complete a demo of your product or service. You can then send unique CTA emails every week or month to keep them engaged and informed.

Giving your visitors something useful, free of charge not only gets their attention but it also gives you a measurable way to determine which leads are genuine and give solid information to your sales team to act on. Consider putting an image of your display in your follow-up email so that your recipient remembers who you are straight away. Keep in mind that they visited many exhibits during the trade-show, and this is a sure-fire way to help them sift through the chaos.

Record Everything

In order to measure your success, you need to document everything. Make sure you have a record of marketing collateral used, pre-event and post-event correspondence and the results it gained. Report on your goals and how close you came to achieving them, or how far you overshot them.

Keep your budget clear. Make sure you know how much you spent on pre-event, during event and post-event marketing and branding. This way, you'll be able to see how well you've done, and if your exhibiting experience was worth it.

You'll be able to quantify whether the leads you gathered gave you a good return of investment. Also, you'll be able to decide which trade shows have worked for you and your brand, and which you should maybe skip the next time around.

There's no point in returning year after year to a trade show that hasn't given you the results that you'd hoped for. Having documentation of everything to do with your trade show will allow you to make tweaks to disappointing campaigns, or overhaul your whole strategy if the results were extremely poor.

Tricks of the Trade Show

The trade show you decide to participate in is bound to have a few companies who do exactly what you do. But there are other exhibitors who have the exact same audience you do but for a different reason. For example; if your company installs fireplaces and there's another exhibitor who installs fire alarms; you'll have the same target audience, but you're selling something different.

In this case trading leads and contact information is a win-win situation. This is a great trade show tip in terms of return on investment. Most trade shows have exhibitors with the same target audience as you. Why not trade leads and contact information? Or go one step further, consider coordinating your post-trade show email campaign. Or even, offer a giveaway together if you can make contact with this other business before the event day.

Your People are Important

Don't forget that staffing at your trade show booth is also important. Having your best salespeople engaging with the show attendees is a guaranteed way to keep the conversation going. It can be one of the deciding factors whether or not an attendee will leave their information with you or not.

If you have to hire promotional staff to man the booth, make sure they've attended a pre-show briefing and that they understand your service or product.

There's nothing worse than visiting an exhibit where a young person sits on their phone and isn't interested in engaging with your visitor. There's no better way to see your leads walk away than this.

Your exhibit staff should be able to speak in depth about the product or services you are offering. They need to be able to treat people with utmost respect and attention, whether they are an employee of your company or not.

Prioritize Your Leads

Leads fall into the following categories:

Cold: Reach out to them at least within the first two or three days after meeting them at the event. The sooner you reach out, the more likely they are to remember you and recall the conversation you had.

Warm: Send an email within the first two days, and consider also calling them after sending the email to try to reach them sooner.

Hot: You can often send these leads to the sales team or through to the later stages of the marketing funnel to further solidify the relationship before beginning negotiations.

When creating your follow-up strategy, personalize each stage of communication based on what kind of lead they are. This can help you dedicate more time to the people who are most likely to want your products, services or partnership while maintaining and building relationships with those who may not be interested or ready to move onto the next steps.

Organize Your Leads Based on Readiness

Before you can understand your company's different levels of prospect readiness, you need to have clear communication about what an ideal, or "quality", lead looks like. Should they have acompany size minimum? An annual revenue range that is ideal? Should they be in a specific industry? Without this clarity, some reps may label leads as "hot" that others would label as "cold". Or worse, they may neglect to follow up with the perfect lead due to a miscommunication. Consistency across the company is critical to capturing accurate information.

At some shows, lead information is captured when you scan their badge and is automatically imported into a portal. It is up to the exhibiting company to update leads in their system for hot, warm and cold statuses.

Your follow-up communication will differ based on how you categorize leads. For example, hot leads need immediate outreach from your sales team, while warm and cold leads can be put in an email nurture sequence until they're ready to make a decision.

Guide Leads Through the Marketing Funnel

Create a lead-nurturing campaign that continues to make meaningful contact with prospective customers until they're ready to purchase. Place hot and warm leads on a communication path with individualized emails that encourage them to take the next steps, like demo the product or service.

As they take each new step, they progress through the funnel, and you can track where each connection is at any time and brainstorm ways to move them along the funnel more efficiently.

Maintain Connections with Non-customers, too

Those who don't fit your target customer profile or wouldn't benefit from your product or service can still be great resources for your organization through future partnerships or just spreading the word about your brand. Add another segment to your marketing funnel that motivates and informs these connections. That way, when you and your team find an opportunity, these connections might be suited for, they're still familiar with the organization, which can save you time when trying to move them to the next steps.

Planning ahead and creating your trade show follow-up strategy in advance of the trade show will help make post-show follow-up a breeze. Take the above tips into consideration prior to your next trade show to be sure you're equipped with a well thought out follow-up strategy for the best ROI possible.

(This article is edited according to Internet information)

 NEW WORDS AND PHRASES

access   n. 机会

showcase   vt. 展示, 展现

close the sale   完成销售, 成交

leading up to   在……之前

in this day and age   在当今时代

harness   vt. 控制并利用

campaign   n. 有计划的活动

action   vt. 执行, 实施

no point   毫无意义

hordes of   成群结队的

gamify   vt. 游戏化

stop in   中途作短暂停留

entice   vt. 引诱

up to date with   与……保持同步

enthusiastic   adj. 热情的, 热心的

verification   n. 核查

in the long run   从长远来看

cover   v. 涵盖

takeaway   n. 要点

competition   n. 对手, 竞争者

a special off   特别优惠价

sign up for   报名参加

demo   n. 演示, 展示

solid information   可靠的消息/资料

straight way   立刻, 马上

sure-fire   adj. 一定成功的

sift though   筛选

document   vt. 记载, 记录

marketing collateral   市场营销材料

skip around   跳过去

make tweak   作出微调

overhaul   vt. 全面改革

briefing   n. 情况介绍会

marketing funnel   营销漏斗

breeze   n. 轻而易举的事

well-thought-out   adj. 慎重考虑而产生的, 深思熟虑的

 NOTES

1. Salesforce, 客户关系管理（CRM）软件服务提供商。其总部设在美国旧金山, 可提供随需应用的客户关系管理平台。其是全球按需 CRM 解决方案的领导者, 提供按需定制的软件服务, 用户每月需要支付类似租金的费用来使用网站的各种服务, 这些服务涉及客户关系管理的各个方面, 从普通的联系人管理、产品目录到订单管理、机会管理、营销管理等。

2. A man who does not plan long ahead will find trouble at his door. 人无远虑, 必有近忧。

3. marketing funnel, 营销漏斗模型。是指在营销过程中, 将非用户（潜在客户）逐步转变为用户（客户）的转化量化模型。营销漏斗的关键要素包括营销的环节和相邻环节的转化率。营销漏斗模型的价值在于其量化了营销过程中各个环节的销量, 以帮我们找到薄弱环节。营销漏斗模型不是固定的, 但其最终结果一般是相同的, 就是达到用户购买或消费的目的。

 EXERCISES

Ⅰ. Answer the following questions.

1. There are definitive marketing techniques for trade shows that one should follow before the event. What are these marketing techniques?

2. What is the great idea to make that first contact and potentially turn your lead into a customer?

3. When you are going to send post-event emails, what suggestions are given to you?

4. Please list some tricks of the trade show.

5. Why do you maintain connections with non-customers?

Ⅱ. Please translate the following English sentences into Chinese.

1. We also suggest creating a timeline for yourself so that you can stay on schedule and don't miss any important details of the planning process.

2. Before you can action a post-show strategy and follow up, you'll need to gather some information from your booth visitors.

3. There's no point in placing a brochure rack right at the entrance to your exhibition stand.

4. If your end goal is an email follow-up campaign, then obviously your goal during the trade show is to collect email addresses and other contact information.

5. Even a day or two after the show will set you miles ahead of your competition who may take weeks to send out their first line of post-show communication.

6. Post-event email is a great way to make that first contact and potentially turn your lead into a customer.

7. Consider putting the name of the show in the subject line, and even personalize your email campaign.

8. Make sure you have a record of marketing collateral used, pre-event and post-event correspondence and the results it gained.

9. Having documentation of everything to do with your trade show will allow you to make tweaks to disappointing campaigns, or overhaul your whole strategy if the results were extremely poor.

10. Planning ahead and creating your trade show follow-up strategy in advance of the trade show will help make post-show follow-up a breeze.

Ⅲ. There are 10 sentences in this section. Beneath each sentence there are four words or phrases marked A, B, C and D. Choose one word or phrase that best completes the sentence.

1. Which of the following is the best translation of the phrase "return on investment"? (　　　)

A. 投资恢复　　　　B. 投资回报　　　　C. 投资回报率　　　　D. 投资收益

2. Please choose the best one among the following Chinese translations for "A man who does not plan long ahead will find trouble at his door". (　　　)

A. 人无远虑, 必有近忧

B. 不作长远打算的人会自找麻烦

C. 一个不计划好未来的人必然有眼前的麻烦

D. 不做计划的人注定要失败

3. What is best Chinese translation of the phrase "tricks or the trade show"? (　　　)

A. 诡计　　　　　　B. 骗局　　　　　　C. 窍门　　　　　　D. 把戏

4. Leads usually fall into (　　　) categories?

A. 4　　　　　　　B. 5　　　　　　　C. 3　　　　　　　D. 2

5. The phrase "prospective customers" is also called (　　　).

A. future customers　　　　　　　　B. expected customers

C. potential customers　　　　　　　D. due customers

6. Before you can <u>action</u> a post−show strategy and follow up, you'll need to gather some infor-mation from your booth visitors. What does "action" mean in this sentence? (   )

A. sue   B. litigate   C. carry out   D. process

7. A trade show gives you direct <u>access</u> to these key decision makers. What does "access" mean in this sentence? (   )

A. approach

B. a code to get use of sth.

C. opportunity

D. operation of reading or writing stored information

8. In order to measure your success, you need to <u>document</u> everything. Which of the following has the closest meaning with "document"? (   )

A. record   B. paper   C. evidence   D. a file

9. Perform a Game or Do a <u>Giveaway</u>. Which of the following has the closest meaning with "giveaway"? (   )

A. free gift   B. leakage   C. let out   D. game show

10. Having your best salespeople <u>engaging with</u> the show attendees is a guaranteed way to keep the conversation going. Which of the following has the closest meaning with "engaging with"? (   )

A. communicating with   B. employing

C. taking part in   D. fighting with

IV. Writing

You are the marketing manager of Sunshine Import & Export Co. Ltd. Your company has signed up for International Builders' Show which is to be held in Las Vegas from Jan. 31, to Feb. 2, 2023. Write an IBS Pre−planning to describe your preparation before IBS. (About 150−200 words)

Reading 2

Prompting and Responding to Sales Enquiries

Prompting and responding to sales enquiries is defined as building procedures to answer en-quiries and initiate contact with prospective buyers. There are several ways to reach targeted cus-tomers: trade fairs and exhibitions, buyers and sellers' meetings, promotional web pages, personal sales visits, direct sales literature, trade missions, press releases in the trade press, sales adver-tisements in newspapers are some forms. This task is about selecting them.

Independently of the instrument used to contact clients, a crucial issue for a successful trans-action is the allocation of responsibilities for contacting clients. Making sales contacts may be left into the hands of few who do not have a well−designed standard procedure to follow. This results in arbitrary offers, varied promises and unexpected compromises. Similarly, there may be no clear procedures for specifying who should respond to inquiries and how. This may result in delays in re-sponse, wrong responses and even lost sales. Development of detailed procedures (e. g., how to respond to a phone call) and training of staff is important.

Trade fairs and exhibitions are efficient ways for reaching target clients to generate sales enquiries. Trade fairs and exhibitions are basically a meeting point for suppliers, sellers and investors. This allows managers to have immediate access to information about the market and the competitors. In addition, it provides access to new markets as well as a window to advertise the company's products.

Once the decision to participate in a particular trade fair or exhibition is taken, there are still a number of important issues the manager should consider to ensure a successful participation. Two of the most important issues are the preparation of the stand that will act as a "selling" point, and the implementation of a marketing strategy to implement during the fair.

Trade missions share some similarities with trade fairs but are significantly different in a number of important aspects. Trade missions could have several simultaneous goals, such as the search of possible joint ventures, the search for new suppliers, the improvement of relations with existing clients and the generation of new relations with potential ones, etc.

An important issue for the manager is to determine whether the company will embark in a trade mission by itself or if, on the contrary, it will rely on the help and assistance of trade associations that prepare trade missions for their members. Since by definition a trade mission requires the manager (together with other people) to travel outside the region where its activities take place, it is necessary to have a complete knowledge of how to operate and solve practical problems in the place of destination.

Another way to reach clients is by providing stores with Point of Purchase (POP) displays (for instance near cash registers) that give the store the extra inventory it needs to cope with high volumes of sale, and provides the product with the extra visibility to achieve impulse buying.

The business could also benefit from the use of the Information and Communication Technologies (ICT) to reach clients. ICT allow for a great deal of flexibility and adaptability, while keeping costs relatively low. Among the most used ICT it is possible to find the use of web pages and e-mails. ICT can sometimes be a double edge sword because if a web page is not designed properly (e. g., keeping in mind the needs and requirements of clients) it will not generate the sales expected and it could also discourage the client from requesting further information. The e-mail has to be also used very carefully: the message has to be clear and the language has to be appropriate.

The choice about which method, or combination of methods, to use depends on which is the most cost effective one. In order to do this the business should have implemented systems to track the effectiveness of these methods. If the business has done this well, it will have generated sales leads and enquiries from prospective customers requesting a sales offer. In other words, once the manager knows what he/she has to do (content, above) he/she has to think on how to allocate the resources efficiently and effectively.

Regardless of the method to contact the client selected by the business, human recourses are crucial in the success of this task. The necessary personnel have to be available and ready to plan, prepare and make the most of every contact with an actual or potential client.

Managers will have to make sure all the necessary systems and procedures are in place to: a plan, e. g. , determines what needs to be done, what the objectives of those actions are, how will those things be done, who will execute them and when; Execute, e. g. , obtain and allocate the necessary resources; Control, e. g. , develop checkpoints and procedures to know when things are done, evaluate if they are done appropriately, assess if and how they can be done better in the future.

Pricing and quoting are tasks that concern determining the final or tentative selling price of goods or services and preparing quotations. The enterprise should have in place standard procedures for pricing and quoting. To achieve these, it is essential that the personnel responsible for pricing and quoting be well trained.

Pricing is an extremely critical task for an enterprise and a very powerful strategic variable. Pricing is usually discussed at two different points in time: during the strategy cycle to decide whether the price would be a "high" or "low" (positioning) and during the transaction cycle when the final offer is given to the client.

Competitive pricing means that a lot more must be considered in addition to costs. Costing is related to the number of resources (expressed in monetary units) spent in producing a product under given manufacturing conditions. Pricing is related to the amount of money received by selling that product under existing market conditions. In a sense they are two separate exercises, as one does not necessarily dictate the other. Eventually, however, the process of manufacturing and marketing must end in a profit, therefore pricing and costing in reality do influence one another.

Cost analysis is very important to develop this task. Sufficient information on costs must be available to make accurate estimates of the cost of producing an offer. The main components of total cost are the cost of production, the cost of distribution and the cost of marketing (it is possible to include also the costs associated with the different payment mechanisms and schemes).

Market research is fundamental for successful pricing. When calculating prices the enterprise should take into consideration the buyers' business, supply and demand, the role of competitors and other aspects such as profit margins. The enterprise should monitor prices for competitive and substitute goods.

The ultimate arbiter of the chosen price is the customer. The way the price is set is a vital component of the extended positioning of the product as well as the enterprise's ability to attract and retain customers.

There are many pricing methods enterprises can use; only some are presented here:

Cost plus: the price is computed as a profit margin added to the estimated cost of the product. This system is widely used by SMEs. Cost plus pricing often diminishes cost discipline and reduces the competitiveness of the enterprise.

Market penetration: A low enough price to stimulate demand for a new product and reach as many buyers as possible.

Price differentiation: When different groups of buyers have different willingness to pay for very similar products. In this case the optimal policy consists in setting prices according to consumer's demand elasticity.

Market segmentation: Similar to the previous but offering the same product to different groups of consumers. For this pricing strategy to be successful the business must be able to separate the different groups.

Capital recovery period: When the market is not expected to last for long the adequate price policy should attempt to recover the investment in the shortest period of time.

Promotional prices: To temporarily set low prices in order to encourage consumers to buy the product. This is very useful when the business is introducing a new product. Pricing should take into account the life cycle of the product.

Pricing based on rates of return: The prices of the products are set in such a way that a pre-determined rate of return for the investment is attained.

Pricing based on consumer's perceived value: The price is based upon consumer's perceived value of the product. Production costs are just a lower bound for price setting.

Pricing based on competitors: This method consists in mimicking the prices that competitors have previously determined.

A quotation is a formal presentation of the price and the terms of sale under which the seller is willing to make the sale. The quotation represents an offer that if accepted by the buyer is a binding commitment to both parts. When preparing quotes, the priority of the enterprise should be to obtain orders that will be profitable to it in the immediate or long run rather than undercutting its competitors' prices.

Procedures for preparing quotes must be clear. The quotes themselves must be clear and contain all the relevant details requested by the customer. Therefore, the person in charge of preparing the quotes needs to clarify any details of the customer's requirements which are not yet clear and respond appropriately. In addition to containing information about price, terms of delivery, etc., the price quotation is also a selling document, and must be accompanied by all necessary sales information.

Quotations may include information items from the following list without being limited to it: timing of the typical buying seasons, nature of trial orders (e. g. product ranges or quantities), payment methods (how to identify and decide on the best method of payment balancing theobjectives of the customer, the bank, and the business' own, improving cash flow and minimizing the risks of non-payment), credit periods, delivery terms, patterns of product distribution, accepted mark-ups, product characteristics, packaging methods and buyer expectations of the exporter regarding promotional support.

It is especially important to state clearly the terms of sale, the payment conditions, and the responsibilities of each part and the validity of the offer. It is advisable to include in the quotation any health, safety, quality or environmental assurance standards achieved by the enterprise (e. g., ISO 9000). This is especially important when any of these standards is mandatory in the target market.

The choice of payment methods, insurance, credit, transport, etc., affects the quote and provides elements for negotiating. Accepted industry practices vary in different parts of the world and in different markets. Local knowledge should be taken into account when making quotations.

The most important components of the quotations concern business tasks. The choice of payment method concerns especially the managerial activity of "maintaining an adequate level of working capital". This activity consists in assuring that the enterprise has adequate cash at all times, and it primarily deals with managing funds temporarily invested in inventories (including work-in-process) and in accounts.

(Edited from The Tasks of Business Management System by ITC)

 NEW WORDS AND PHRASES

enquiry   n. (贸易) 询盘

initiate   vt. 开始实施, 发起

prospective   adj. 潜在的, 有希望的

allocation   n. 分配, 配置

arbitrary   adj. 任意的, 随心所欲的

inquiry   n. 询盘, 询价

investor   n. 投资者

implementation   n. 实现, 履行

simultaneous   adj. 同时的, 同时发生的

embark   v. 从事, 着手

point of purchase   购买点

register   v. 登记, 注册

inventory   n. 存货, 清单

cope with   处理, 应付

Information and Communication Technology   信息通信技术

sword   n. 刀, 剑

generate   vt. 生成, 产生

allocate   vt. 分配, 拨出

execute   vt. 实行, 执行

checkpoint   n. 检查站, 检查点

tentative   adj. 试探性的, 暂定的

quotation   n. 报价, 报价单

positioning   n. 定位, 布置   v. 定位(position 的现在分词), 放置

monetary   adj. 货币的, 财政的

manufacturing   adj. 制造的, 制造业的

sufficient   adj. 足够的, 充分的

mechanism   n. 机制

scheme   n. 计划, 方案, 机制

calculate   vt. 计算

profit margin   利润率, 边际利润率

monitor   vt. 监控

arbiter   n. 仲裁者, 裁决人

diminish   v. 减少, 缩小

optimal policy　最优策略,最佳策略

elasticity　n. 弹性,弹力,灵活性

capital　n. 资金,资本

temporarily　adv. 临时地,临时

perceive　vt. 察觉,感觉,理解;认知

bound　n. 限制,界限

priority　n. 优先事项,最重要的事

binding　adj. 有约束力的,必须遵守的

quote　v. 报价　n. 报价,报价单

trial order　试订单

cash flow　现金流

minimize　vt. 使减到最少到最低限度,最小化

mandatory　adj. 强制性的,义务的

 **Reading 3**

Preparing the Export Order for Distribution

Preparing the export order for distribution is defined in terms of scheduling freight operations to achieve distribution objectives, as well as arranging and facilitating the physical movement of goods from the enterprise to its customers. For exporters this task is further complicated due to documentation requirements and multi-modal transport requirements. Most exporters use freight forwarders for this purpose or their customers take possession of the goods at the factory gate (e. g. , Ex-Factory). Even if the buyers receive the goods at the factory gate, the enterprise must carefully plan and monitor the movement of goods to maximize cost savings.

Once the order is produced, international transport arrangements must be made. It is also necessary to pack and mark the goods, and insure them so that they arrive safely at the buyer's preferred location. The amount of export documents required at this stage of the export transaction is considerable. Companies must have qualified personnel who know everything about freight documentation. The enterprise needs to be organized so that documents are available in the right form and shape when needed. The enterprise needs to make sure that freight personnel (employees charged with moving the products from the enterprise to its customers) have formal skills training.

Distributing products internationally brings added complications and risks to their delivery. The passage of goods across national boundaries means more legal requirements, such as customs clearance and compliance with the laws of the importing country. A longer time in transit means that the goods are exposed to the risk of theft, damage and delays, particularly if the goods are being transferred from one mode of transport to another. In addition, the enterprise may have to pack goods differently to meet the size and weight restrictions imposed by different forms of international transport.

Costs of physical distribution can constitute a significant portion of total costs. To be competitive, the seller must continually develop and use a physical distribution system that requires the least cost to achieve the greatest customer satisfaction in terms of the criteria of quantity, quality, time and cost. Freight classifications used by the enterprise should be well researched. The shipping performance must be evaluated and improved according to these criteria. The costs of each task must be evaluated, both in terms of preparing the order and shipping the order.

Preparing the export order for distribution involves the following important activities:

Packing is concerned primarily with protecting and containing the product during transit while packaging is concerned primarily with presenting the product so that it appeals to the customer at the time of purchase. The type of packing used will depend upon a number of factors like: product characteristics, mode of transport, climatic conditions, and government regulations.

Marking consists in identifying the goods en route to the customer. Marks on an export shipment identify the goods to the exporter, the transport carrier, customs officials and the importer. The purpose of these marks is to aid identification, ensure that the goods arrive safely at the point of destination, and comply with government and contractual obligations.

Labelling provides information on the quantity and quality of the goods and also may include the name and address of the manufacturer, the weight or volume of contents, ingredients and other relevant details.

Inspection concerns the verification measures undertaken to ensure that the export order in question meets the exporter's government standards, specific standards demanded by the customer in the terms of sale, legal requirements of the importing government. Inspection requirements can differ enormously from country to country and from product to product, they involve a range of governmental, quasi-governmental and private inspection services.

Unitization describes the way small items of cargo are put together and handled as a unit of standard size, usually using mechanical equipment. The consolidation of an export order into units of a convenient form, weight or volume for transportation can offer distinct advantages to the exporter and the importer. Big savings can be made on shipping costs by choosing the right orientation of goods in transit.

Clearing customs requires the exporter to be aware of export documents and procedures. These are domestic customs procedures, transit procedures and import customs procedures. Exporters can use customs brokers; customs clearing agents or rely on their importers to handle the customs clearance at the point of destination. Sometimes the freight forwarder is authorized to act as a customs clearing agent. In this case, his or her charges should be passed to the customer.

Cargo insurance is used against damage. The exporter must know the nature of insurance required, how to get it, what it covers and how it relates to the terms of sale. Insurance coverage can be negotiated with a broker to make sure that risks specific to the company's own circumstances are included.

Terms of sale are also relevant. The INCOTERM chosen will determine which portions of the delivery correspond to the buyer or to the seller and thus will determine, within the context of the current order, the content of the task being discussed.

(Edited from The Tasks of Business Management System by ITC)

NEW WORDS AND PHRASES

multi-modaltransport   多式联运

freight forwarder   货运代理, 货运代理人

transaction   n. 交易, 买卖, 业务

custom clearance   海关放行, 清关

in term of   在……方面; 依据, 按照

appeal   vi. 有吸引力   n. 吸引力, 感染力

en route   (法) 在途中

obligation   n. 义务, 责任

enormously   adv. 非常地, 在极大程度上

quasi-governmental   准政府的, 半官方的

unitization   n. 单元化

custom broker   报关行

custom clearing agent   通关代理

 ENGLISH FOR WORKPLACE COMMUNICATION

Sample Dialogue: Gillian discusses a trade show with Christopher, who will attend.

Gillian: Hello, Christopher. I need to speak to you. You're going to the Boston Trade Show, aren't you?

Christopher: Am I? That's news to me.

Gillian: Oh, I thought you already knew. Steve can't go, his wife is expecting a baby, so it has to be you.

Christopher: My first trade show, I'm quite excited.

Gillian: There's a lot to plan and think about. Take some notes.

Christopher: Okay, let me get a pen, it's next Thursday, right?

Gillian: Yes, the trade show opens on Thursday 18th and will run for two days. It finishes at 6pm Friday. Now, have a look at this plan, there's the floor plan for the whole thing.

Christopher: Wow, it's big.

Gillian: Yes, it is. Our stand is number three hundred and twenty-one.

Christopher: That's a good position on the corner.

Gillian: Yes, I think we should do very well in this trade show. A lot of the big companies, Google, Microsoft, Amazon will be in that half of the room, so I think we're going to get a lot of passing traffic.

Christopher: Where will our competition be?

Gillian: Good question. Console Solutions are in 322, almost opposite you, so you'll be able to keep an eye on them. DMZ are in 124 on the other side of the hall. I'm not sure about SysNet.

Christopher: They didn't go last year, did they?

Gillian: No, I know they're attending Tech Expo in San Diego the following week, so maybe they won't be at this trade show. I hope not! Right, Christopher, all your hotel and flight details have been arranged. I think you'll get an e-mail with all of that tonight. You'll be traveling with two others: Graham Nash and Liz Sorel.

Christopher: Alright.

Gillian: Here's what we need you to do before departure. Read the trade show manual, that will help you to understand everything about the show.

Christopher: Where can I find that?

Gillian: I have the PDF on my laptop, I'll e-mail it to you. You need to set targets for what you want to do at this trade show.

Christopher: You mean, how many software packages we can sell?

Gillian: Yes, but other things too. How many business cards you can give out, how many journalists you can speak to, how many phone numbers from educational institutions you can collect, that type of thing.

Christopher: Of course, these trade shows are about becoming better known.

Gillian: Exactly! Go and speak to Ray Jones. He's the guy who designed our stand for the trade show. He'll be in his office at three, he's expecting you. The stand is quite high tech with some awesome software demos, but you need to learn how to use it!

Christopher: Ray Jones. Okay, I'll go and see him. Anything else?

Gillian: Yes, we need to change our website. We need to put something about our participation at the Boston Trade Show, advertise our attendance, what do you think?

Christopher: I'll speak to the website group and ask them to put something on the top of every page.

Gillian: Yes, something like that. Tomorrow, I'll be calling all our big customers and telling them that we'll be in Boston next week.

Christopher: It's going to be great.

Gillian: You'll enjoy it, Chris. You'll be tired, but you'll enjoy it.

UNIT 5

GLOBAL EXHIBITION INDUSTRY

本章导读

随着经济全球化程度的日益加深，会展业已发展成为新兴的现代服务型产业，成为衡量一个城市国际化程度和经济发展水平的重要标准之一。根据全球展览业协会（UFI）的数据，全球每年约有 1 200 个展览场馆和 31 000 场展览会。它们提供的展览面积超过 3 480 万平方米，相当于 460 个足球场。每年约有 440 万名参展商和 2.6 亿多观众聚集在展会上。展览会可提供 231 000 个工作岗位。

本章列举了德国、美国、英国三个主要国家会展业发展情况。通过本章的学习，了解会展业对经济的拉动作用，熟悉会展业数据统计的基本方面。国际会展业能增加不同地域、不同文化背景、不同传统习俗的人们之间的互相交流与了解，消除沟通障碍，扩大共识，为产品的跨区域、跨文化、跨民族、跨环节的流通创造条件，有利于供给实现和供给创造。

知识精讲

 展会知识

德国探索会展经济转型

近日，德国国家旅游局、欧洲活动中心协会和德国会议促进局联合发布报告，对德国 2021 年会展市场的关键数据进行了分析，认为会展行业在新冠肺炎疫情影响下正积极谋求转型，线上虚拟会展、线上线下相结合的混合会展以及可持续性会展将成新趋势。

会展业是德国经济的重要组成部分。据数据门户网站 Statista 的统计，截至 2020 年，德国共有近 7 600 家会展中心、会议酒店等场所。德国伊福经济研究所的统计显示，来自全球各地的参展商和参观者每年在德国会展市场的支出达 145 亿欧元，创造 23.1 万个工作岗位。

疫情对德国会展行业造成了较大冲击。仅 2020 年一年，超过 70% 的贸易展会被迫取消或延期，全年营业额比预期减少约七成。

为缓解会展行业遭受的损失，2021 年 1 月，总额为 6.42 亿欧元的"德国联邦拯救伞计划"获批，德国会展行业相关企业可申请补贴，对象包括会展设施所有者、经营者及中

介机构，最高补贴额为其全部利润损失。此外，德国经济事务与气候行动部联合各联邦州，推出了总额为 6 亿欧元的保险项目，以帮助展会组织者应对疫情带来的次生灾害。

与此同时，德国会展行业积极探索新的办展方式。德国展览业协会表示，展会组织者加速数字化转型，以维持与客户的联系，并提供新产品信息。《会议和活动晴雨表2021—2022》报告显示，2021 年，混合展会数量增长了 280%，线上虚拟展会活动增加120%。在参与人数方面，2021 年，参加混合展会活动的人数为 1 840 万，而在 2020 年只有 180 万人。

德国会议促进局局长马蒂亚斯·舒尔策认为，参加会展活动的方式不断变化，反映了商业会展活动日渐多元化的趋势。德国会展活动组织者已经认识到线上虚拟展会和混合展会的优势。同时，很多人对于真实环境下面对面交流的愿望，也正随着会展行业的新变化作出相应的调整。"德国作为全球会展活动重要目的地之一，已经为这种多样化需求做好了充分准备。"舒尔策说。

（《人民日报》2022 年 7 月 5 日）

思考并回答以下问题：

1. 德国主要的会展城市都有哪些？德国知名的会展公司有哪些？这些公司每年举办的主要展览会都有哪些？

2. 以德国科隆五金展为例，分析该展览会为何成为全球五金行业的风向标？

3. 疫情下，全球会展业遭受重创，德国会展公司做了哪些改变？这些变化是否有成效？

4. 讨论：线下展览会与线上展览会的区别有哪些？

5. 调查：分组分别调查至少 20 家外贸企业，设计一份问卷，谈谈他们对线下展会和线上展会的感受，问卷可涉及以下几个方面：预算、展示方式、客户来源、产品体验、安全等。

Reading 1

German Exhibition−Number One Worldwide

Trade fairs are a key driver of international trade in goods and services. They are important marketing instruments in B2B communications and they intensify competition and trade in nearly all economic regions around the globe, ensuring growth and jobs. Economic globalization and an increased orientation towards brand names are additional factors promoting the worldwide importance of these sector−specific marketplaces. In the process, trade fair organizers are developing into marketing partners for businesses in ever more comprehensive ways.

The trade fair spectrum covers everything from special exhibitions for nearly every sector in highly developed countries to universal exhibitions in developing countries. According to data from the Global Association of the Exhibition Industry (UFI), there are around 1, 200 exhibition venues and 31, 000 exhibitions a year worldwide. They offer a combined hall space of more than 34.8 million square meters-or the equivalent of 460 football fields. Around 4.4 million exhibitors and more than 260 million visitors gather at these more than 31, 000 exhibitions a year. Germany accounts for 10 percent of the world's trade fair market.

The functions of trade fairs

Of all marketing instruments, trade fairs have by far the broadest range of functions. They directly affect business administration, national economies and society as a whole.

MARKETING FUNCTIONS

Trade fairs serve to establish and maintain customer relations, find business partners and personnel, and position companies as a whole. As test markets for new products, they are also market research instruments. They enable companies to raise name recognition levels, analyze competitive environments and prepare to sell products and services. Just visiting trade fairs can pay off for young companies, for example in the early stages of entering markets.

OVERALL ECONOMIC FUNCTIONS

Trade fairs benefit their exhibitors and visitors, but also the economies around their host cities and even their respective states and countries. They generate strong secondary effects, especially for the hotel, restaurant and transport sectors. They are also very beneficial to companies that provide valuable services to trade fair organizers and exhibitors, in fields such as stand design and construction, logistics, translation and interpreting, and hosting. Regional economic effects can be five to sevenfold the sales for organizers, especially if the exhibition venues have a strong international focus. Trade fairs therefore ensure many jobs in their regions, particularly for medium-sized companies.

SOCIAL FUNCTIONS

Trade fairs have always been centers of knowledge of information that is prepared, cultivated and placed in helpful contexts. As our society becomes ever more knowledge-based, information has become a crucial resource. Producing, selecting, filtering and channeling it has thereby become one of the most important activities of national economies. As a result, ever more conferences are being held in conjunction with exhibitions, and vice versa, as vibrant and immediate ways of conveying knowledge.

Number one worldwide

Germany is the world's number one location for international trade fairs. Some 160 to 180 international and national trade fairs are held in the country every year, with around 180,000 exhibitors and ten million visitors. Trade fairs in Germany bring partners together from around the world. They are forums for communication and innovation that reflect the world's markets. Around two-thirds of all global trade fairs are held in Germany.

Important for the overall economy

Exhibitors and visitors spend a total of around €14,5 billion a year for their activities at trade fairs in Germany. The overall effect on economic production amounts to €28 billion German exhibition organizers post sales of around €4 billion a year. Of the ten highest-grossing trade fair companies in the world, five are headquartered in Germany.

Trade fair organization secures a total of about 231,000 jobs. An average of two employees at exhibiting companies work on trade fairs. With around 58,000 companies currently active in the B2B segment at trade fairs, that means more than 100,000 full-time jobs.

Advantages for Germany as a trade fair location

STATE-OF-THE-ART EVENT FACILITIES

Germany has 25 exhibition venues of international or national significance, with a combined hall space of 2.8 million square meters. The country's exhibition facilities set international standards in architecture, logistics and technology. German exhibition centers invest around 300 million euros a year in optimizing their facilities.

Four of the world's eight largest exhibition venues are located in Germany, and ten venues in the country each have hall capacities of more than 100,000 square meters. Regional exhibition centers offer an additional combined hall space of around 380,000 square meters.

MARKET-ORIENTED TRADE FAIR STRATEGIES

German trade fairs address long-term economic needs and reflect the importance and innovative power of their respective sectors. Close cooperation among organizers, exhibitors and visitors ensures long-term, market-oriented strategies and ideal trade fair dates.

HIGH INTERNATIONAL PRESENCE

A special competitive advantage of German trade fairs is their international appeal-the fairs draw the world's markets into the country. Almost 60% of the approximately 180,000 exhibitors a year come from abroad, and one-third of these from countries outside Europe. Of the 10 million visitors each year, nearly 30% come from abroad.

HIGH PROFESSIONALISM

German trade fair organizers are professionals in the field of international exhibitions. Some of these companies are responsible for holding more than 20 of the world's leading trade fairs a year. Their highly specialized experts with international experience are responsible for every aspect of the trade fair business.

LEADING SERVICE STANDARDS

German organizers offer exhibiting companies a wide range of services. They support exhibitors by booking travel and accommodations and by doing press, publicity and marketing work. They also continuously expand their spheres of activity. In addition, many trade fair organizers have installed permanent online marketplaces, making them expert marketing partners for exhibiting companies throughout the year.

EXCELLENT COST-BENEFIT RATIO

Trade fairs in Germany have moderate stand fees compared to other international sites. At the same time, they attract a high quantity and quality of visitors, which means that exhibitors in the country come into contact with many potential customers. Costs per visitor contact are favorable compared to trade fairs in other countries and to other marketing media.

ATTRACTIVE REGIONAL TRADE FAIRS

The international and national trade fairs are supplemented by a dense network of well-organized regional trade fairs for distinct target groups. They include both specialist and consumer fairs. These events draw a total of around 50,000 exhibitors and 6 million visitors a year.

The exhibition business structure in Germany

ORGANISERS

Approximately 100 exhibition organizers are active in Germany, around 40 of which handle international fairs. The largest of them are among the highest-grossing trade fair companies in the world. This makes the exhibition business one of the leading service sectors in Germany, also in comparison to other countries.

The German organizers in AUMA put on around 300 trade fairs in other countries-primarily in major growth regions such as Asia, North America, South America and Eastern Europe. That too benefits the German economy, because it needs expert partners for its trade fair activities-particularly in highly competitive foreign markets.

EXHIBITING COMPANIES

Around 58,000 German companies are active exhibitors in the B2B segment. The majority of them are in the manufacturing sector (55%), followed by the service sector (23%), and trade (20%). Medium-sized companies dominate in terms of both number of employees and sales: 51% of exhibitors have fewer than 50 employees, and 39% have 50 to 499. Some 47% of exhibiting companies post sales of up to 2.5 million euros, and 35% of 2.5 million to 50 million euros.

VISITORS

The percentage of decision-makers among all trade fair visitors is exceptionally high at 63%. Managing directors, board members and self-employed people from Germany make up 35% of trade visitors, and 73% of those from abroad. The latter group commands above-average decisional powers, with 91% having determining or co-determining influence on business decisions. Thirteen percent of trade fair visitors come from companies with more than 1,000 employees, including top decision-makers from global corporations. Some 54% of trade visitors come from companies with fewer than 60 employees.

SERVICE COMPANIES

Close cooperation with a large number of service companies is what makes a trade fair successful. Of special note here are stand constructors, designers, event specialists and consulting companies, as well as shipping companies, stand personnel trainers, caterers and hotels. The stand construction companies in the FAMAB professional association alone post overall annual sales of around €2 billion.

German trade fairs abroad (GTQ)-At home on the world's markets

If you and your company are seeking to enter new markets in other countries, you need strong local partners. German trade fairs abroad open doors for you and facilitate key contacts to international trade partners.

AUMA and FAMA members organize more than 300 trade fairs a year in major growth regions outside Germany, especially in Asia, North America, South America and Eastern Europe-and this number is growing. Taking interval-based fluctuations into account, the number of visitors at German trade fairs abroad is rising at a nearly continuous rate.

Successful dual strategy

German trade fair organizers offer their event and sector-related expertise not only in their home country but also around the world. Their event strategies are based on those of established international trade fairs in Germany. This enables you as an exhibitor to enter markets far from home, with the assurance of German quality standards under the label "German Trade Fair Quality Abroad". In addition, they develop new trade fair topics that are tailored precisely to specific regions.

DUAL STRATEGY OF GERMAN TRADE FAIR ORGANISERS

German trade fair organizers pursue three main aims with their dual strategy:

They demonstrate expertise in selected sectors worldwide;

They attract new exhibitors and visitors for their major fairs in Germany;

They benefit from high rates of growth outside Europe;

By putting on an increasing number of trade fairs abroad, German organizers seek to attract the corresponding comprehensive international involvement (exhibitors and visitors) at their major fairs in Germany.

At the same time, German organizers are also demonstrating their expertise in selected sectors worldwide. At locations abroad they can reach companies that would not participate in trade fairs in Western Europe. Moreover, the quality of their trade fairs can then also convince these companies of the merits of travelling to Germany to exhibit at major trade fairs in their sectors there.

Superior services for exhibitors

At German trade fairs abroad, you and your company can concentrate on what you do best-taking care of your customers and your business. Organizers from Germany will handle everything else, which benefits you as an exhibitor in numerous ways. You can rely on their long years of experience in putting on international trade fairs in specific sectors. You can also benefit from their contacts to business associations both in Germany and the target countries, as well as to potential customers in these locations.

German organizers usually also have subsidiaries in regions where they work, which are superbly networked within those countries. They can open doors to customers in areas that have not yet come into contact with companies from abroad.

GERMAN TRADE FAIR QUALITY ABROAD-THE GTQ LABEL

Together with its members that are active abroad, AUMA has developed the German Trade Fair Quality Abroad (GTQ) label. It highlights the strong quality standards of trade fairs organized by German companies in other countries. To use the GTQ label for their trade fairs abroad, AUMA and FAMA members must meet these four criteria:

They own the basic concept;

They are the main overseer for putting the concept into practice;

They create the budget;

They run international sales;

The organizers publish figures on their stand surface areas, number of exhibitors and number of visitors for their foreign trade fairs under the GTQ label.

Every autumn AUMA publishes a calendar of dates for the year to come. It can be downloaded as a PDF file or ordered in printed form: German Trade Fair Quality Abroad 2019.

Exhibition Industry-Key figures

Trade fairs are by far the most important instrument in B2B communications. This is one of the main reasons why the exhibition business is one of the leading service sectors in Germany. It encompasses all companies, associations and persons that organize fairs or provide services for organizers and exhibitors. The sector also enjoys a high status internationally.

Review 2021

After 2020, Germany as a trade fair venue had to contend with massive economic slumps in 2021 as well, due to the Corona pandemic. Of the 380 planned trade fairs, only 100 could be held in the third and fourth quarter. At the international, national, and regional trade fairs held in Germany, 1. 3 million square meters of exhibitor stand space, 36,000 exhibitors and 2.1 million visitors were counted. Comparing 2021 with an average year before the Corona pandemic, there were around 85 percent fewer exhibitors and 87 percent fewer visitors at German trade fairs. The exhibitor stand space rented showed a deficit of 85 percent. This means that 2021 was even worse than the already appallingly bad year of 2020 (Figure 5. 1 and Figure 5. 2).

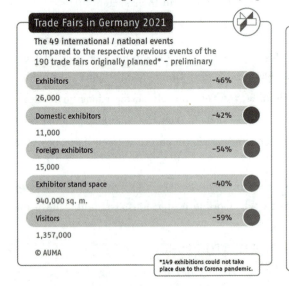

Figure 5. 1   **Figure 5. 2**

The global exhibition market

In international comparison, too, the German trade fair industry is highly esteemed. Four of theten largest trade fair companies in the world are based in Germany. In the international turnover ranking, five of the top 10 trade fair organizers are from Germany–measured by turn over before the Corona pandemic. According to the international trade fair association UFI, there are around

1, 360 trade fair venues worldwide and 31, 000 trade fairs per year. Germany's share of surface area is eight percent.

REVENUES OF EXHIBITION COMPANIES WORLDWIDE (Figure 5. 3)

Figure 5. 3

HALL CAPACITIES WORLDWIDE (Figure 5. 4)

Figure 5. 4

The German exhibition market

Twenty–five venues for international or national trade fairs in Germany have a combined hall space of around 2.8 million square meters. Ten sites each have capacities of more than 100, 000 square meters, and five additional sites each have more than 50,000 square meters. Hall space isexpected to grow only slightly in the coming years. Investments will be made primarily in quality. An additional approximately 390,000 m^2 of hall space are available at venues for exhibitions with regional focuses.

THE GERMAN EXHIBITION INDUSTRY—TURNOVER OF THE EXHIBITION COMPANIES

According to the organizers' plans, 370 to 380 fairs were to be held in 2020 and 2021. More than 70 percent of the fairs had to be cancelled or postponed due to Corona. All in all, the German trade fair organizers recorded sales losses of 65 to 70 percent in each year respectively (Figure 5.5).

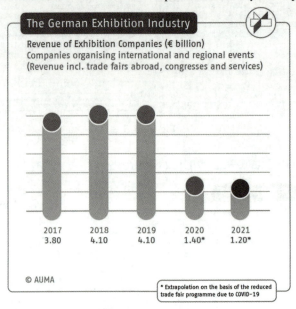

Figure 5.5

HALL CAPACITIES IN GERMANY ACCORDING TO VENUE (Figure 5.6)

Trade Fairs in Germany 2022

Exhibition capacities*
gross in sq. m.

Location	Halls	Outdoor	Location	Halls	Outdoor
Hanover	392,445	58,000	Dortmund	63,000	80,000
Frankfurt/M	372,350	66,764	Karlruhe	60,000	90,000
Cologne	285,000	100,000	Augsburg	54,500	10,000
Dusseldorf	262,727	43,000	Bremen	39,000	100,000
Munich (Exh. Center)	200,000	414,000	Erfurt	25,070	21,600
Berlin ExpoCenter City	190,000	157,000	Offenburg	22,500	37,877
Nuremberg	180,000	50,000	Freiburg	21,500	81,000
Stuttgart	119,800	40,000	Offenbach	20,100	
Leipzig	111,300	70,000	Chemnitz (Exh. Center)	11,000	8,000
Essen	110,000	20,000	Wiesbaden	10,000	
Friedrichshafen	87,500	35,500	Husum	4,800	70,000
Hamburg	86,465	10,000	Idar-Oberstein	4,500	900
Bad Salzuflen	78,000				

© AUMA

* Locations with at least one event according to AUMA category international and national events / Status: April 2022

Figure 5.6

THE GERMAN TRADE FAIR SECTOR-OVERALL ECONOMIC IMPORTANCE (Figure 5. 7)

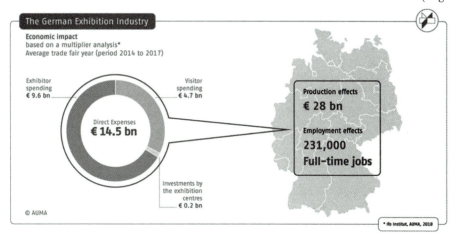

Figure 5. 7

International and national trade fairs in Germany

After 2020, the international and national trade fairs in Germany also had to contend with massive economic slumps in 2021 due to the Corona pandemic. Of the 198 planned fairs in this category, only 49 could be held in the third and fourth quarter (Figure 5. 8) .

Trade Fairs in Germany 2021

The 49 international / national events in comparison to the respective previous events (in %)

	Number of events	Exhibitors	Stand space	Visitors
Total	49	−49.0	−40.0	−50.0
Investment goods fairs	35	−44.9	−46.1	−50.5
Consumer goods fairs directed to trade visitors	7	−59.4	−38.6	−68.0
Consumer goods fairs directed to the public	6	−19.0	−20.9	−41.5
Trade fairs presenting services	1	−42.9	−36.7	−60.0

© AUMA

Figure 5. 8

Trade fairs in Germany-Development 2015−2019

Nearly two−thirds of the leading global trade fairs in different branches of industry are held in Germany. The country is the world's number one location for international trade shows. Despite the worsening economic conditions, they again grew slightly in Germany compared to their previous e-vents, with only the number of visitors remaining constant. The competitive position of trade fairs vis−à−vis other marketing media remains stable.

TRADE FAIRS IN GERMANY-ABSOLUTE NUMBER OF EXHIBITORS (Figure 5.9)

Figure 5.9

TRADE FAIRS IN GERMANY-ABSOLUTE NUMBER OF VISITORS (Figure 5.10)

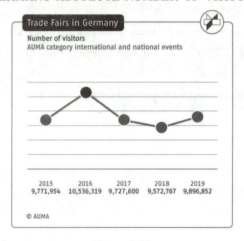

Figure 5.10

TRADE FAIRS IN GERMANY-ABSOLUTE EXHIBITOR STAND SPACE (Figure 5.11)

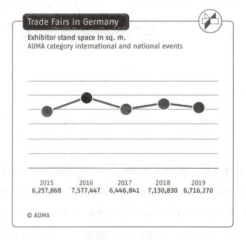

Figure 5.11

Participation from abroad

Internationality on the exhibitor and visitor side is the most important advantage of German trade fairs in global competition. For many years, the decisive growth driver in international participation on the part of exhibitors was the region of South−East−Central Asia. Almost 31,000 exhibitors (+11.9%) from Asian countries came to the trade fairs in Germany in 2019. The number of exhibitors from EU countries fell by 2%.

TRADE FAIRS IN GERMANY-ABSOLUTE NUMBER OF FOREIGN EXHIBITORS (Figure 5.12)

Figure 5.12

TRADE FAIRS IN GERMANY IN 2019-FOREIGN EXHIBITORS BY REGIONS (Figure 5.13)

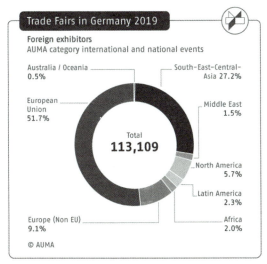

Figure 5.13

TRADE FAIRS IN GERMANY IN 2019-MOST IMPORTANT EXHIBITING COUNTRIES (Figure 5.14)

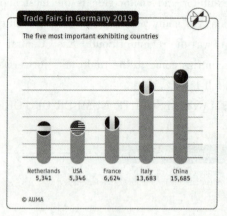

Figure 5.14

TRADE FAIRS IN GERMANY-ABSOLUTE NUMBER OF FOREIGN VISITORS (Figure 5.15)

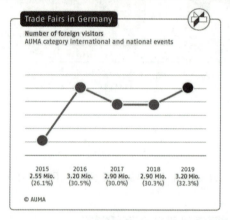

Figure 5.15

GERMANY AS A TRADE FAIR LOCATION IN 2019-FOREIGN VISITORS BY REGIONS (Figure 5.16)

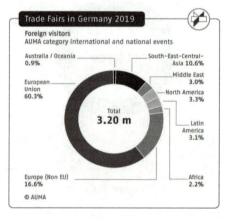

Figure 5.16

TRADE FAIRS IN GERMANY IN 2019-MOST IMPORTANT VISITOR COUNTRIES (Figure 5. 17)

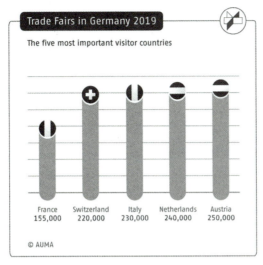

Figure 5. 17

Regional trade fairs

International and national trade fairs in Germany are supplemented by a dense network of regional trade fairs for both trade and public visitors. Some 55, 000 exhibitors and around 6 million visitors gather at these events every year.

Key figures 2021

The 52 trade fairs with a regional catchment area recorded a total of almost 10, 000 exhibitors, 374, 000 square meters of stand space and 764, 000 visitors in 2021. Originally, 181 fairs in this category were planned, but 129 were cancelled or postponed due to the pandemic (Figure 5. 18) .

Regional Exhibitions 2021

The 52 regional exhibitions in comparison to the respective previous events (in %)

	Number of events	Exhibitors	Stand space	Visitors
Total	52	-40.0	-30.0	-60.0
Special-Interest-Public Exhibitions	29	-34.5	-25.3	-52.4
Multi-Sector-Public Exhibitions	10	-56.1	-52.0	-68.6
Trade Fairs	13	-23.2	-17.9	-29.1

© AUMA

Figure 5. 18

German trade fair activity abroad

In addition to participating in trade fairs at home, German businesses also make considerable use of trade fairs abroad. German companies participate in an average of ten fairs every two years: six in Germany and four in other countries. A good third (36 percent) of these companies selected trade fair locations in Europe in 2017/2018, and 24 percent overseas. These figures are from the 2019 AUMA MesseTrend, a representative survey (carried out by TNS Emnid) of 500 German exhibiting companies (Figure 5.19).

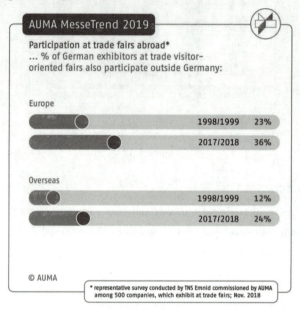

Figure 5. 19

AUMA MesseTrend 2020 survey-Graphics

A few months before the Corona pandemic, in November 2019, 500 representatively selected German companies that primarily exhibit at business-to-business fairs were surveyed. The result once again impressively showed the fundamentally strong anchoring of the trade fair as an instrument in German companies. According to the survey, 29% of exhibitors wanted to spend more money on trade fair participation in 2020/2021, 53% wanted to invest about the same amount, and only 17% planned to spend less. These were the findings of the AUMA MesseTrend 2020, a survey conducted by Kantar TNS on behalf of AUMA. But even in this survey it became clear that real experiences and virtual reality will complement each other in the future-at least on exhibition stands of German exhibitors. For the time of the Corona pandemic and beyond, a considerable strengthening of this development is expected (Figure 5.20–Figure 5.22).

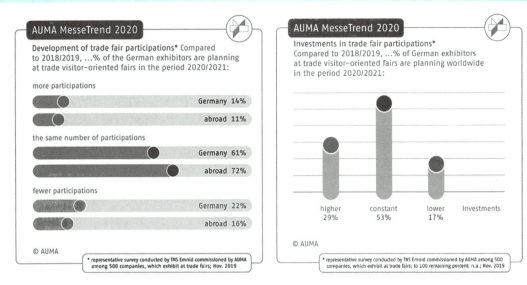

Figure 5. 20　　　　　　　**Figure 5. 21**

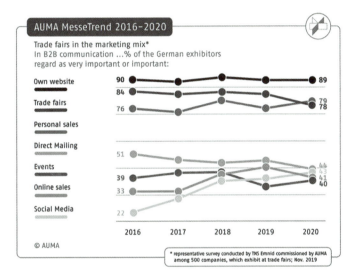

Figure 5. 22

Trade fairs as central marketing instruments

Successful companies have clear goals and pursue them with an equally clear strategy. On their road to success, they use a mixture of marketing tools. In this interplay of suitable marketing instruments, trade fair participations by far the widest range of functions—from acquiring new customers to forging press contacts. Digital extensions also ensure that even more benefits can be generated by increasing the reach. It therefore only makes sense that German companies make intensive use of trade fairs.

 NEW WORDS AND PHRASES

venue　　n. 聚会地点,场馆

vibrant　　adj. 充满生机的,鲜艳的,强劲的

headquarter   n. 总部

logistic   n. 物流，后勤

approximately   adv. 接近地，大约地

tailor   n. 裁缝，定做

corresponding   adj. 对应的，相应的

superbly   adv. 极好地，庄重地

slump   n. 下跌，衰退

esteem   vt. 尊重

interplay   n. 相互作用

intensive   adj. 广泛的

mixture   n. 混合物

complement   n. / vt. 补充，不足

survey   n. 调查

fundamentally   adv. 基础地，基本地

impressively   adv. 令人印象深刻地

representative   adj. 代表意义的

primarily   adv. 首要地

appallingly   adv. 骇人听闻地，令人惊讶地

encompass   vt. 围绕，包围

dual   adj. 双重的

expertise   n. 专门技能，专业知识

exceptionally   adj. 罕见地，特别地，例外地

Euro   n. 欧元

install   vt. 安装

accommodation   n. 住宿

optimize   vt. 最大化

sevenfold   n. 七倍

equivalent   adj. 等同的，相等的

promote   vt. 提升，促进

 NOTES

1. Global Association of the Exhibition Industry（UFI），全球展览业协会，是国际展览协会最重要的组织之一，创立于 1925 年，总部设于巴黎。

2. virtual reality，虚拟现实。

3. Corona pandemic，新冠疫情。

4. the GTQ，德国海外贸易展。

5. AUMA，德国经济展览和博览会委员会。德国经济展览和博览会委员会在联邦经济与技术部和消费者保护、营养与农业部协助下，为德国官方参与国外展会计划做筹划准备工作。

6. pay off，偿还、还清。

7. secondary effect，次发效应、间接效应。

8. the leading service sector，服务行业中的领先部分。

9. the third and fourth quarter，第三和第四季度。

 EXERCISES

Ⅰ. Answer the following questions.

1. What are the social functions of trade fair?

2. What are the advantages for Germany as a trade fair location?

3. Briefly explain the exhibition bushiness structure in Germany.

4. What is the development of trade fair in Germany during 2015 to 2019?

5. Briefly explain the service companies in Germany.

Ⅱ. Please translate the following English sentences into Chinese.

1. In the process, trade fair organizers are developing into marketing partners for businesses in ever more comprehensive ways.

2. The trade fair spectrum covers everything from special exhibitions for nearly every sector in highly developed countries to universal exhibitions in developing countries.

3. Germany accounts for 10 percent of the world's trade fair market.

4. Of all marketing instruments, trade fairs have by far the broadest range of functions.

5. As test markets for new products, they are also market research instruments.

6. As our society becomes ever more knowledge−based, information has become a crucial resource.

7. They are forums for communication and innovation that reflect the world's markets.

8. A special competitive advantage of German trade fairs is their international appeal the fairs draw the world's markets into the country.

9. Some of these companies are responsible for holding more than 20 of the world's leading trade fairs a year.

10. In addition, they develop new trade fair topics that are tailored precisely to specific regions.

Ⅲ. There are 10 sentences in this section. Beneath each sentence there are four words or phrases marked A, B, C and D. Choose one word or phrase that best completes the sentence.

1. Of all marketing instruments, what by far has the broadest range of functions? (   )

A. companies   B. trade fairs   C. services   D. market research

2. What share does Germany account for in the world's trade fair market? (   )

A. more than 20 percent   B. less than 5 percent

C. about 10 percent   D. almost 30 percent

3. What would trade fairs benefit? (   )

A. exhibitors   B. visitors   C. host cities   D. all the above

4. Which country is the world's number one location for international trade fairs? (   )

A. Germany   B. United States   C. France   D. China

5. What does the label GTQ stand for? (　　)

A. General Trading Quantity　　　　　B. German Trading Quality

C. Genetic Transform Question　　　　D. German Trade Fair Quality Abroad

6. Why is mainly responsible for the massive economic slumps in 2021? (　　)

A. overproduction　　　　　　　　　B. poor management

C. Corona pandemic　　　　　　　　　D. communication barriers

7. What does turnovers of the exhibition mean? (　　)

A. increases in application　　　　　　B. cancellation or postponing

C. presence in exhibition　　　　　　　D. attending other exhibition

8. For many years, which of the following region is the decisive growth driver in international participation on the part of exhibitors? (　　)

A. African countries　　　　　　　　　B. European countries

C. South-East-Central Asia　　　　　　D. North American

9. What is the role of information as a resource? (　　)

A. crucial　　　　B. irrelevant　　　　C. harmful　　　　D. Detrimental

10. According to the deficit of exhibitor stand space, compared with 2020, what is the situation of 2021? (　　)

A. slightly better　　B. much better　　C. even worse　　D. the same

Ⅳ. Writing

If you were an executive running exhibition stand space, facing frequent turnovers recently, please write a short report:

1. to summarize what happened recently.

2. to list finding details in your work.

3. to make possible recommendations.

Reading 2

Exhibition Venues in the United States

The Las Vegas Convention Center

The Las Vegas Convention Center is one of the busiest facilities in the world a 4.6 million-square-foot facility located within a short distance of approximately 150,000 guest rooms. Operated by the Las Vegas Convention and Visitors Authority (LVCVA), the center is well known among industry professionals for its versatility. In addition to approximately 2.5 million square feet of exhibit space, 225 meeting rooms (more than 390,000 square feet) handle seating capacities ranging from 20 to 2,500. Two grand lobby and registration areas, one located in the West Hall, and the other in Central Hall (more than 260,000 square feet) efficiently link existing exhibit halls with new exhibit and meeting rooms, and allow simultaneous set-up, break-down and exhibiting of multiple events.

The LVCC was awarded the Global Biorisk Advisory Council (GBAC) STAR facility accreditation by ISSA, the world's leading trade association for the cleaning industry. Considered the gold standard for safe facilities, the GBAC program was designed to control the risks associated with in-fectious agents, including the virus responsible for COVID-19. The LVCC was the first facility in Nevada to receive the accreditation.

In the 1950s, the Las Vegas city and county leaders recognized the need for a convention fa-cility. The initial goal was to increase the occupancy rates of hotels during low tourist months. Leaders chose a site one block east of the Las Vegas Strip at the site of the Las Vegas Park Speed-way, a failed horse and automobile racing facility from the early 1950s. A 6,300 capacity, silver-domed rotunda with an adjoining 90,000 square feet exhibition hall opened in April 1959. It hos-ted The Beatles on August 20, 1964.

In 2018, the Las Vegas Convention Center released plans to undergo yet another $890 mil-lion expansion, the 14th in its history. The expansion intended to increase the center's meeting space and improve the building's overall design. Updates would feature the latest in technology, as well as to connect the Convention Center to the Las Vegas Strip. The authority has announced plans to expand the direction of the LVCC by creating a Las Vegas Global Business District. Those plans resulted in the announcement for the acquisition of the Riviera in February 2015 for $182.5 million.

World Market Center Las Vegas

World Market Center Las Vegas, located at 495 Grand Central Parkway in Las Vegas, Ne-vada, is a 5-million-square-foot (460,000 square meters) showcase for the home and hospitality contract furnishings industry in downtown Las Vegas. It is the largest showroom complex in the world for the home and hospitality furnishings industry, serving domestic and international sellers and buyers.

World Market Center is owned by International Market Centers, L. P. (IMC). On May 3, 2011, Bain Capital, Oaktree Capital Management, The Related Companies, and other investors announced the formation of International Market Centers, a venture that combines properties in High Point and Las Vegas and would control 11.5 million square feet of showroom space. At that point, two of the World Market Center buildings and two of the High Point showroom complexes had been in receivership for several months. In a deal completed September 26, 2017, Blackstone Real Estate Partners, Blackstone Tactical Opportunities and Fireside Investments purchased IMC, which at the time owned 5.4 million square feet in three Las Vegas locations.

In August 2019, construction began on the Expo at World Market Center building. It is a 315,000 square feet (29,300 m^2) convention facility that will connect to Building C. The expo building will replace the Pavilion tent facilities located across the street. The tents would be removed upon completion of the expo building and the 21-acre land will initially be used for parking, with the potential for future development. The expo building was topped off on December 19, 2019, and was opened on April 9, 2021.

When not hosting the Las Vegas Market or other trade shows, the first two floors of Building A are open as showrooms for interior designers and the general public Monday-Friday 10am-5pm. These spaces are called the Las Vegas Design Center (LVDC) and contain more than 400,000 square feet (37,000 m^2) of permanent showrooms, which are open Monday-Friday. The LVDC offers year−round access to a global selection of furniture, fabrics, lighting, floor coverings, wall décor and accessories.

Part of Las Vegas Design Center's ongoing programming includes the First Friday series, a monthly A&D event featuring top design and arts and cultural speakers from around the world. The Design Icon award is part of the Design Series and celebrates modern−day design legends and gives them a platform to share their stories and inspire others.

Chicago McCormick Place

McCormick Place is the largest convention center in North America. It consists of four inter-connected buildings and one indoor arena sited on and near the shore of Lake Michigan, about 2 miles (3.2 km) south of downtown Chicago, Illinois, United States. McCormick Place hosts numerous trade shows and meetings. The largest regular events are the Chicago Auto Show each February, the International Home and Housewares Show each March and the National Restaurant Association Annual Show each May and the International Manufacturing Technology Show in the fall every other year.

As early as 1927, Robert R. McCormick, a prominent member of the McCormick family of McCormick Reaper fame, and publisher of the Chicago Tribune, championed a purpose−built lakeside convention center for Chicago. In 1958, ground was broken for a $35 million facility that opened in November 1960, and was named after McCormick, who died in 1955. The lead architect was Alfred Shaw, one of the architects of the Merchandise Mart. This building included the Arie Crown Theater, designed by Edward Durell Stone. It seated nearly 5,000 people and was the second largest theater (by seating capacity) in Chicago.

The original McCormick Place, completed in 1960, seen in 1966 from Lake Michigan before it's destruction by fire in 1967. Although many wanted to rebuild the hall on a different site, Chicago mayor Richard J. Daley elected to rebuild on the foundations of the burned building.

On January 3, 1971, the replacement building, later called the East Building and now called the Lakeside Center, opened with a 300,000 square feet (28,000 m^2) main exhibition hall. The Arie Crown Theatre sustained only minor damage in the 1967 fire, and so was incorporated into the interior of the new building. The theater, with the largest seating capacity of any active theater in Chicago (the Uptown Theatre having more seating, but currently closed), underwent major modifications in 1997 which improved its acoustics.

The North Building, located west of Lake Shore Drive and completed in 1986, is connected to the East Building by an enclosed pedestrian bridge. In contrast to the dark, flat profile of the East Building, the North Building is white (as the original building was), with twelve concrete pylons on the roof which support the roof using 72 cables. The HVAC system for the building is incorporated into the pylons and give the building the appearance of a rigged sailing ship. The North Building has approximately 600,000 square feet (56,000 m^2) of main exhibition space.

The South Building, dedicated in 1997, contains more than 1 million square feet (93,000 m^2) of exhibition space. It more than doubled the space in the complex and made McCormick Place the largest convention center in the nation. The South Building was built on the former site of the McCormick Inn, a 25−story, 619−room hotel built in 1973 as part of the McCormick City complex and demolished in 1993.

On August 2, 2007, McCormick Place officials opened yet another addition to the complex, the West Building, costing ＄882 million and completed eight months ahead of schedule. The publicly financed West Building contains 470,000 square feet (44,000 m^2) of exhibit space, bringing McCormick Place's total existing exhibition space to 2.67 million square feet (248,000 m^2). The West Building also has 250,000 square feet (23,000 m^2) of meeting space, including 61 meeting rooms, as well as a 100,000 square feet (9,300 m^2) ballroom, the size of a football field and one of the largest ballrooms in the world.

McCormick Place continued to expand in October 2017 with the opening of Wintrust Arena, a 10,387-seat arena situated on Cermak Road just north of the West Building. The new facility hosts DePaul Blue Demons men's and women's college basketball, and the WNBA's Chicago Sky. The new arena boasts 22 suites, 479 club seats, and 2 VIP lounges. The arena is also equipped to host concerts, sporting events, meetings, and conventions in conjunction with the rest of the McCormick Place complex. Sporting events such as gymnastics and volleyball are also held in the McCormick Place buildings in addition to the arena.

The Metra Electric Line stops at a station in the basement of McCormick Place. The South Shore Line also stops at the same station on weekends. The Chicago Transit Authority serves the facility with its Cermak-McCormick Place station on the Green Line, approximately 12 mile west, and two bus routes.

On March 27, 2020, the United States Army Corps of Engineers announced that the complex would begin transforming convention space into a 3,000-bed hospital in the wake of the COVID-19 crisis affecting the Chicago area. The ＄15 million project was paid for by FEMA (Federal Emergency Management Agency) and was scheduled for completion on April 30.

 NEW WORDS AND PHRASES

The Las Vegas Convention Center　拉斯维加斯会议中心

square　n. 平方

versatility　n. 多功能,多才艺

lobby　n. 大堂,大厅

set-up　n. 机构,组织

accreditation　n. 达到标准,证明合格

infectious　adj. 传染性的,有感染力的

virus　n. 病毒

adjoin　v. 紧邻,相邻

showroom　n. 陈列室,样品室

hospitality　n. 热情好客

domestic　adj. 国内的

furnishing　n. 装饰品

properties　n. 特征

completion   n. 完成交易

interior   adj. 内部的

fabric   n. 纺织物, 纤维

floor covering   地毯, 地面覆盖物

accessory   n. 配件, 附属物

champion   n. 冠军

incorporate   vt. 包含, 包括

modification   n. 修改, 改进

acoustic   n. 音响效果

pedestrian   n. 行人, 步行者

pylon   n. 电缆

rigged sailing ship   装配式帆船

ballroom   n. 舞厅

FEMA (Federal Emergency Management Agency)   n. 联邦应急管理署

Reading 3

Global Exhibitions Day: Industry Needs to Promote its Positive Impact on UK Trade

As the exhibition industry celebrates Global Exhibitions Day, a key focus for the NEC Birmingham in the coming months will be to promote the role of exhibitions in supporting business growth across domestic and international markets.

"We need to highlight the volume of trade which takes place at exhibitions, " said Ian Taylor, Managing Director, Conventions & Exhibitions, NEC Group. "In terms of homegrown UK businesses, hundreds of millions of pounds are exchanged. Our experience of the pandemic taught us a valuable lesson; we haven't explained the business impact of exhibitions to the public in the way we need to. We are a physical marketplace, a positive force for UK PLC."

Pre-pandemic, exhibitions generated £11 billion of business sales, contributing £5.4 billion of GDP and supporting 114, 000 jobs. As the exhibition industry recovers, a key focus for the NEC Birmingham will be to promote the opportunities for businesses in connecting with trading partners.

Business Secretary Kwasi Kwarteng said: "I commend NEC Birmingham, and the whole of the UK exhibition industry, for their tireless work putting great British businesses rightfully in the spotlight. The UK is a world leader across industries from aerospace and health technology, to finance and IT, and telling those success stories to the world helps businesses to grow and create jobs across the country.

Next week, Nineteen Group launch Manufacturing & Engineering Week at the NEC Birmingham, developed closely with sector partners, including Make UK. The exhibition will be opened by Business Secretary Kwasi Kwarteng.

"I look forward to flying the flag for the UK's fantastic manufacturers at Manufacturing and Engineering Week at NEC Birmingham, which will be a prime example of how the exhibition industry supports British businesses, large and small, " said the Business Secretary.

Manufacturing and Engineering week takes place 7 – 10 June, bringing together a series of events that will showcase end-to-end manufacturing and engineering solutions.

"Whether you're a developer offering the latest products in your field or a company looking for that extra advantage that world beating technology can offer, Manufacturing and Engineering Week is for you, " said Verity Noon, Marketing Director, Nineteen Group.

"Covering the full life cycle from design, to engineering, to manufacturing and maintenance, the event will bring so many innovative ideas and creative people together, it will help supercharge UK industry to meet its challenges and make the most of its opportunities."

ECONOMIC IMPACT OF EXHIBITIONS IN THE UNITED KINGDOM

What qualifies an exhibition?

UFI follows the ISO 25639-1: 2008 (E/F) definitions which are also adopted here. For the purposes of this study, an exhibition, show, or fair is an event in which products, services, or information are displayed and disseminated. Exhibitions differ from conferences, conventions or seminars, or other business and consumer events. Exhibitions exclude flea markets and street markets. Exhibitions include:

• Trade exhibitions: exhibitions that promote trade and commerce and are attended primarily by trade visitors. A trade exhibition can be opened to the public at specific times.

• Public exhibitions: exhibitions open primarily to general public visitors. A public exhibition is sometimes also known as a consumer show.

In 2018, approximately 1, 100 exhibitions sold more than 4.7 million net square meters in the UK. Exhibitions generated approximately £5.0 (€5.6) billion of direct spending, by visitors, exhibitors and additional exhibitions related expenditure. Exhibitions welcomed nearly 9.1 million visitors and 178, 000 exhibitors to the UK in 2018.

Based on a total of 178, 000 exhibitors in the UK in 2018, direct spending per exhibitor amounted to £27, 899 (€31, 538). Based on a total of 0.67 million square meters of venue capacity measured in terms of gross indoor exhibition space (as reported in UFI's World Map of Venues), direct spending per square meter of venue capacity amounted to £7, 413 (€8, 380).

COMPONENTS OF ECONOMIC IMPACT ANALYSIS

There are three main components of a sector's overall economic impact:

Direct impacts consist of the direct spending and jobs that are involved in planning and producing exhibitions, and for participants to travel to exhibitions, as well as other exhibitions-related spending.

Given the characteristics of the exhibitions sector, much of this direct activity occurs across a variety of sectors. For example, the production of an exhibition frequently involves employees on-site at a hotel or other venue, including banquet staff as well as audio-visual/staging and technical staff, and other third party contracted service providers, such as entertainment/production services, décor, speakers and trainers, advertising and promotion. These employees all represent direct jobs supported by the exhibitions sector. Meanwhile, participants' travel to the exhibition, and accommodation during the event, supports direct spending and jobs across a range of service providers in the travel sector. Though this spending is occurring across businesses in a range of industry sectors, it all represents activity that is supported by exhibitions direct spending, and is part of the exhibition sector's direct impacts.

Indirect impacts represent downstream supplier industry impacts, also referred to as supply chain impacts. For example, the facilities at which exhibitions occur require inputs such as energy and food ingredients. Also, many exhibition venues contract with specialized service providers, such as marketing, equipment upkeep, cleaning, technology support, accounting, and legal and financial services. These are examples of indirect impacts.

Induced impacts occur as employees spend their wages and salaries in the broader economy. For example, as hotel employees spend money on rent, transportation, food and beverage, and entertainment.

Indirect and induced impacts may also be referred to collectively as indirect effects. To conduct the impact analysis, we used country-level economic impact multipliers from the existing exhibitions impact studies. For countries where exhibitions impact multipliers were either unavailable or appeared inconsistent with reference data, we used travel and tourism multipliers maintained by WTTC (World Travel and Tourism Council) and Oxford Economics. WTTC multipliers are based on input-output tables for each country and were sourced from either the OECD (Organization for Economic Co-operation and Development), or when not available, national statistical offices. From the input-output tables, multiplier matrices were developed for each economy, detailing the flow of spending in an economy that occurs as a consequence of spending in a given industry Overall, the total economic impact of the exhibition industry in the UK in 2018 is summarized as follows:

- £11.0 (€12.5) billion of economic output (business sales);
- £5.4 (€6.1) billion in total GDP contribution;
- Nearly 114,000 total jobs.

These totals represent the combination of direct impacts within the exhibitions sector (e. g. £5.0 billion of exhibitions directs pending, and 55,000 direct jobs), plus the estimated indirect and induced effects. The resulting output multiplier for the exhibitions sector in the UK is 2.23, implying that each £1.00 (€1.00) in direct exhibition spending generates an additional £1.23 (€1.23) in indirect and induced expenditures in the UK economy.

 NEW WORDS AND PHRASES

homegrown   adj. 本土生长的
physical marketplace   实体市场
recover   v. 恢复
commend   vt. 要求
rightfully   adv. 正当地
spotlight   n. 聚光灯, 公众注意
tireless   adj. 不知疲倦, 精力充沛的

aerospace   n. 航空航天技术

fantastic   adj. 好极了

showcase   n. 玻璃陈列柜

maintenance   n. 维持, 维系

supercharge   n. 增压器

conference   n. 大会, 峰会

flea   n. 跳蚤

net   adj. 净(产值, 增长等)

expenditure   n. 开支, 费用

downstream   adj. 顺流的, 在下游

ingredient   n. 成分, 要素

upkeep   n. 保养, 维修

inconsistent   adj. 不一致的

multiplier   n. 乘法, 加算器

expenditure   n. 花费, 费用

 ## ENGLISH FOR WORKPLACE COMMUNICATION

Sample Dialogue: Office Life

Steve: Do you know who is going to Brussels for the product launch?

Helen: Barry and Susan, I think.

Steve: Barry needs to stay here to prepare next year's budget. William told me Charlie would be going.

Helen: Oh, now I remember. It's Susan who has to prepare the budget so it must be Barry he'll accompany.

Helen: Did you hear we didn't get the ComTel contract?

Steve: I don't believe it. What on earth happened?

Helen: They told us that the costing of the factory construction was too expensive.

Steve: I suppose Harbon Industries won it, did they?

Helen: Yes, they were willing to guarantee that the factory would be built in Egypt and that swung the deal.

Steve: Do you have any idea how this works?

Helen: Switch it on using the red button. Wait until the green light shows, which means it's hot enough. Then slide the document in and it comes out the other side.

Steve: I need to use a plastic sleeve, don't I?

Helen: Yes, put the document inside it.

Helen: Tomorrow is Monday the fourth, isn't it? That's the deadline for the Spain project. We'll never get it finished in time.

Steve: What are you worried about? The deadline isn't until Friday. That's the eighth.

Helen: Steve, what are you talking about? Didn't you get the memo changing the deadline? I thought we were going to cancel the project and, instead, we have to get it done by tomorrow. There's too much to do still: the sales projection, wages, overheads and somebody has to write the proposal.

Steve: Look, I spoke to Mr. Gonzlez yesterday and he said it would be fine to wait until the end of the week. We've got nearly a week.

Steve: Did you go to the meeting about relocating next year?

Helen: I did. Some people got a bit hot under the collar.

Steve: How come?

Helen: Well, the management is presenting this as something that is really exciting, but I'm not sure that everybody sees it that way.

Steve: What did you say, Helen?

Helen: I said that meetings like that are a huge waste because nothing we say will change anybody's mind. I think we will have to find a new job or move house.

Steve: I'm looking for the file we were working on the last week, you know, the Woodward contract.

Helen: It should be in the folder, "Germany".

Steve: Why do we have a folder called Germany when there is also a folder called Europe. That's the type of confusion which results in lost documents. Are you sure it hasn't been deleted? It's not in the Germany folder.

Helen: I don't know what to say. I wasn't the last one to work on it.

Steve: Oh, look, it's sitting here on the desktop. Someone must have saved it there by mistake.

UNIT 6

WORLD FAMOUSE TRADE FAIRS

本章导读

通过本章的学习，了解几个国际知名展览会，如美国芝加哥国际家庭用品展览会、美国拉斯维加斯国际五金展览会和德国法兰克福国际汽车配件展览会。国际知名展会国际化程度高，参展商和采购商来自全球不同的国家和地区，并在某个行业具有一定的知名度，且通常引领行业发展方向，国际知名展会能为买卖双方创造更多的产品价值、信息价值和交易价值。

了解这些国际展会的发展历史、展出内容、覆盖市场、买家需求等可以帮助外贸企业提供决策和开拓国际市场的咨询服务。一场国际性展会通常会根据市场变化进行不断创新，这也是为了更好地服务于参展商和采购商。

知识精讲

展会知识

2022 年法兰克福汽配展参展商阵容与疫情前一样国际化

2022 年 9 月 13—17 日，全球领先的德国法兰克福汽配展（Automechanika）将在法兰克福会展中心举行。今年展会的参展商阵容与疫情发生前一样国际化，各家汽车制造商和供应商将在现场展示他们的最新产品和解决方案。同时，超过 350 场面向汽车专业人士的活动将分四个阶段举行，主题包括替代驱动系统、培训和专业发展、弹性供应链和电子商务。

"德国是一个汽车大国，汽车行业面临着巨大的挑战，包括大量的软件和硬件技术更新。即将举行的法兰克福汽配展将以研讨会和演讲的形式为这些挑战提供广泛的答案。在本届展会的 2 800 多家参展商中，仅 17% 为德国参展商，这意味着参展商阵容的国际化构成与疫情前旗鼓相当，但是令我们遗憾的是，今年几乎没有企业前来参展。"法兰克福展览公司执行委员会成员 Detlef Braun 总结道。

一个新的展区——"Innovation 4 Mobility"将在 3 号展厅展示互联网汽车的开创性解决方案。互联网汽车、自动驾驶和替代能源等主题将由奥迪、ADOBE、博世、波士顿咨询集团、eBay、谷歌云、KEYOU、弗劳恩霍夫研究所、舍弗勒和丰田汽车基金会组织。

电子商务主题也在法兰克福汽配展中占有一席之地。eBay 将首次设立展台，展示如何成功进入线上零售市场以及如何优化现有的电子商务运营。9 月 16 日，amz Talk 将探讨一个问题："汽车零部件售后市场的数字化和生态系统——电子商务和线上市场能否改善采购流程？"

供应链短缺这个紧迫的话题将在首届"Automechanika 供应链管理日"活动上进行详细的探讨，该活动将于 9 月 15 日上午 10：00 至下午 5：00 在 4 号展厅举行，主要针对汽车行业物流和供应链创新解决方案的用户和供应商。一系列有趣的演示和研讨会将为参与者提供充分的机会来获取信息、建立联系和寻找新的合作伙伴。

今年的 Automechanika 创新奖将于 9 月 13 日颁发，它不仅仅提供汽车售后市场令人印象深刻的展示，还将反映当前汽车配件行业的发展趋势。来自 99 家参展商的 133 份参赛作品涵盖了令人印象深刻的应用范围，从可再生资源生产的油漆护理产品、电池起火车辆的救援系统，到能够在几分钟内测量整个车辆的激光扫描仪。颁奖典礼将于展会首日下午 5：00 在法兰克福会议中心举行。

（本文根据互联网新闻报道整理）

思考并回答以下问题：

1. 查询相关资料，了解德国法兰克福汽车配件展览会。

2. 论坛是展览会上重要的活动之一，根据新闻报道，2022 年的法兰克福汽配展都有哪些主题论坛？

3. 创新是所有国际展会的核心，谈谈你对产品创新的理解。

4. Automechanika 作为汽车行业展会的一个品牌，目前在全球举办，你是否知道都在哪些国家和城市举办过？

 **Reading 1**

The Inspired Home Show

The International Home + Housewares Show staged every year since 1906, is organized by the International Housewares Association (IHA). It is the center of the IHA's yearly activities. In 2008, the Show covered 785, 000 net square feet (72, 900 m^2) of exhibit space. It is one of top 20 largest trade shows in the U. S. and in the top 10 in Chicago.

The first House Furnishing Goods Exhibition was held in New York's Madison Square Garden in 1906. The modern housewares exposition was born in 1927 when members of the National Home Furnishings Buyers Club decided that an exhibit in centrally located Chicago would be the most efficient way to view the products of many manufacturers. The group convinced 115 manufacturers to display their wares at the Stevens (later renamed Conrad Hilton) Hotel on January 3-7, 1928, now known as the Hilton Chicago. The newly established National House Furnishing Manufacturers Association (NHFMA) responded to their buyers' request for an annual exhibition, and for the next 10 years, shows including kitchenware and major appliances were held at the Stevens.

IHA was formed from multiple mergers among different organizations. The Housewares Show grew swiftly after World War Ⅱ ended. Exhibitors continually strained to expand their booth space, and would be exhibitors battered at the doors. To accommodate the growing number of companies, the Show moved to Chicago's Navy Pier in 1949.

By 1949 the show was an international marketplace hosting buyers from 11 countries. The Show narrowed its product categories, terminating the major appliance segment and changing its name from the "National Housewares and Major Appliance Exhibit" to the "National Housewares and Home Appliance Manufacturers Exhibit" in 1950. The 1950 event saw a greater number of new products, although many manufacturers still had not invested in new tooling. The Show's primary importance was its abundance of personal contacts.

In 1956 no more than 649 exhibitors could fit into Navy Pier. The limitations could only be solved when the Show moved into Chicago's new exposition center, McCormick Place on the Lake, in 1961.

A new McCormick Place rose, and in January 1971, the 54th International Housewares Show was back at the exposition center. 700,000 square feet (65,000 m^2) replaced the former facility's 480,000 square feet (45,000 m^2). In 1979, the West Building (Donnelley Hall) was pressed into service for new exhibitors with an extra day to attract buyers. McCormick Place's new North building added 300,000 square feet (28,000 m^2) in 1986 to a Show that occupied three buildings.

In 1991, NHMA moved to new quarters in Rosemont near O'Hare Airport. The show changed its title several times, most notably from the "National Housewares Exposition" in the 1980s to the "International Housewares Show" in 1992, when it became a single annual January Show in Chicago. In 1997, the International Housewares Show opened in the grand new South building of the McCormick Place complex. In 2004, IHA moved the trade show from its long-standing January date to a March timeframe and renamed it the International Home + Housewares Show in recognition of show's evolution to a home goods marketplace.

IHA is committed to maximizing the success of the home + housewares products industry. IHA provides the world-class marketplace, The Inspired Home Show, as well as facilitation of global commerce, executive-level member share groups and conferences, a wide range of international business development tools, housewares industry market data and information services, facilitation of industry standards, and more!

For more than 80 years, IHA has hosted the industry's premier home + housewares trade show. Retail buyers come to Chicago every year to discover new products and innovations, to meet with current suppliers and to find new opportunities for partnership.

The Inspired Home Show features thousands of product categories across the entire home + housewares industry. We've made it easy to navigate by segmenting the Show into five Expos grouped by product type and each featuring unique on-floor attractions.

Clean + Contain

A PLACE FOR EVERYTHING & EVERYTHING IN ITS PLACE

Entire retail concepts are built around inventive, stylish ways to care for the home. Organization and storage are no longer hidden away in closets—they're designed to be seen, and consumers will pay a premium for options that fuse order with creativity. The Clean + Contain Expo showcases products that both streamline the home and contribute to its aesthetic, so you can delight your customers with fashion and function.

North Hall—Level 3

Dine + Décor

PERSONALITY FOR THE HOME, PROFITABILITY FOR THE RETAILER

Whether a casual night in or a full-fledged dinner party, consumers cultivate their personal brands through at-home entertaining. That means sparing no expense in preparing and serving inspired food in a carefully curated environment. The Dine + Décor Expo is alive with functional and decorative products that create and serve meals with flair—and encourage your customers to splurge on their signature style.

South Hall—Level 3

Discover + Design

UNIQUE PIECES THAT DEFINE THE HOME & THE RETAILER

You've built a customer base on the ability to offer unique, trend-setting items that they've never seen before and can't find anywhere else. Elements that differentiate them from their peers and you from your competition: The Discover + Design Expo was created to make it easy for you to source high-end, design-forward concepts from across product categories and around the world. This is where the discerning retailer finds their game-changing piece.

North Hall—Level 3

Wired + Well

PRODUCTS THAT ELECTRIFY THE HOME, THE SELF & THE STORE

Technology continues to make the consumer's life easier—and with more ease comes more demand. Your customers will spend real dollars on products that do the work for them—providing function, information, feedback and control. The Wired + Well Expo offers a collection of devices that simplify everything from mixing and blending to floorcare to tracking towards fitness goals—functionality that your customers see as everyday necessities.

Lakeside Center—East Building

International Sourcing

The International Sourcing Expo at The Inspired Home Show features more than 500 housewares exhibitors that offer value-driven products from China and Chinese Taiwan. Additional sourcing opportunities exist at the Show through the Global Design Points displays, which include country pavilions from Brazil, Colombia, Egypt, France, Chinese Hong Kong, India, Indonesia, Japan, the Republic of Korea, Turkey and Thailand.

North Hall—Level 1

Housewares Homecoming Brings CHESS, IBC Global Forum and Credit Group Meeting Together in One Event.

Two critical International Housewares Association educational summits, CHESS: Chief Housewares Executive SuperSession and the International Business Council's Global Forum, will be held in-person this fall along with a new event by the recently formed Housewares Credit Group (HCG). Under the banner Housewares Homecoming, the three events will take place together for the first time, uniting industry alumni, veterans and peers for insights and discussions around the most critical topics facing home + housewares suppliers today.

CHESS and the Global Forum will take place Sept. 22 & 23, 2022 at the Westin Hotel in Rosemont, Ill. , near O'Hare Airport. The HCG meeting will be held Sept. 21 at IHA's offices in Rosemont and with a hybrid virtual option.

CHESS

CHESS is IHA's strategic and networking event for industry leaders. It is designed for chief officers of all IHA member companies and their top decision-makers and features sessions on critical issues affecting housewares, this year including an economic outlook discussing the markets and mid-term elections; e-commerce brand control; consumer valuation; optimizing physical retail; and creating opportunities on Web 3.0.

Other sessions will include the annual Housewares Hot Seat panel featuring housewares supplier chief executives discussing remaining proactive in a reactive marketplace. The panel will include Luke Peters, president & CEO, NewAir, and Jacob Maurer, CEO, The Americas, The Cookware Company. Peter Giannetti, editor-in-chief, HomePage News, will moderate the session. In addition, The NPD Group will offer a post-COVID playbook for retail and the home industry, and market research and analytics specialists CivicScience will discuss macro consumer trends impacting home goods.

Phillip Neuhart, senior vice president & director of market and economic research at First Citizen's Bank Wealth Management, will examine current economic and market conditions and share his outlook. Market-driving topics will include Fed policy, inflation, the labor market, consumer spending, economic growth and the mid-term elections.

CHESS is sponsored by Pattern, CIT, Oracle Netsuite and The Lakewood Group.

IBC Global Forum

The Global Forum is the annual meeting of the IBC, a special interest group of IHA members dedicated to global marketing and sales. Sessions are applicable for suppliers new to the international arena or veteran sales executives seeking growth in their global business. They will feature presentations by global retailers and distributors speaking on their markets and best practices for suppliers selling to their firms.

Among the key presentations, Mexican distributor Tendenciasy Conceptos will offer insights on the home and housewares market in Mexico. Online retailers Mathon France and Westwing Brazil will discuss how successful suppliers work with their firms to reach their consumers and France and Brazil.

IHA's global offices and representatives will present the top opportunities for sales in their markets based on current economic conditions, top retailers, and consumer demand. Office and reps attending will represent Brazil, Canada, Colombia, France, India, Mexico, and the UK. Networking sessions will allow attendees to leverage their peers' knowledge to troubleshoot challenges faced by home + housewares suppliers selling into global markets. In addition, The NPD Group and CHESS sponsor Pattern will lead discussions on global sales insights.

Housewares Credit Group

The Housewares Credit Group is a cooperative effort between IHA and Riemer Plus for IHA member financial executives. Members receive an in-depth look at retailer credit standings to help mitigate risk in their sales portfolios. The Housewares Homecoming event will include a financial analysis led by Reimer Plus of national retailers including At Home, Bed Bath & Beyond, Kohl's, Tuesday Morning, Wayfair and more; group account discussions including best practices and exchange of account information—all strictly adhering to anti-trust laws; and a presentation on bankruptcy by Brent Weisenberg, senior counsel at Lowenstein Sandler, LLP, explaining everything suppliers need to know before a customer files bankruptcy.

Exhibitors Return to the Inspired Home Show 2023 After COVID Break in 2022

Top housewares brands that were unable to participate in The Inspired Home Show 2022 are committing to returning to the Show in 2023, the International Housewares Association announced today. More than 100 exhibitors that sat out the 2022 Show have already submitted space applications for the 2023 Show, which is set for March 4-7 at Chicago's McCormick Place Complex.

"We were one of the brands that sat out of The Inspired Home Show 2022," said Rob Michelson, president and CCO, Bradshaw Home. "Although COVID concerns made our participation impossible this past year, we are very excited and enthusiastic to get back to the Show in 2023 and rebuild those face-to-face engagements that are critical to our success."

"As face-to-face events continue to rebound, we are thankful for the support of our exhibitors and the buying community in keeping The Inspired Home Show the premier housewares marketplace," said Derek Miller, IHA president & CEO. "Our industry is most successful when it comes together, and it's wonderful to see the support from so many companies and brands. IHA will continue to work to make the 2023 Show as meaningful and productive as possible for the benefit of all attendees."

The Inspired Home Show 2023 will take place March 4-7 at McCormick Place in Chicago. Buyer registration is currently open at The Inspired Home Show. com/Register. Companies interested in exhibiting at the Show can find more information at The Inspired Homes Show. com/Exhibit.

(This article is edited according to Internet information)

NEW WORDS AND PHRASES

houseware   n. 家居产品, 家用器皿

furnishing goods   家具

kitchenware   n. 厨房用具

appliance   n. 电器, 器皿

merger   n. 机构合并, 企业归并

swiftly   adv. 迅速地

strain   v. 拉拽, 尽力拓展

would-be exhibitor   潜在展商, 未来展商

batter   v. 敲打

terminate   v. 终止

tooling   n. 机床

world-class   n. 世界一流,世界顶级

maximize   vt. 最大化

premier   adj. 首要的,著名的

premium   n. 保险费,附加费

flair   n. 天赋,天资

splurge   v. 挥霍,花钱

signature style   独特风格

floorcare   n. 地板护理

pavilion   n. 设置的席位或看台

analytics   n. 分析学,逻辑分析法

veteran   n. 退伍老兵,有丰富经验的老手

leverage   v. 借贷收购;n. 杠杆效应

troubleshoot   v. 检修,解决困难

mitigate   vt. 减轻,缓和

attendee   n. 参与者

portfolio   n. 文件夹,公文包

rebind   v. 恢复,复苏

NOTES

1. Navy Pier,海军码头。一般指芝加哥市海军码头,位于美国伊利诺伊州芝加哥市密歇根湖边。是一个多功能旅游场所,也是芝加哥的娱乐和会展等活动首选地之一。

2. McCormick Palace,麦考密克会展中心。其是北美最重要的大型会展中心,由三个具有最新水平的展馆构成,分别是南馆、北馆和湖畔中心。这些展馆总展览面积达 220 多万平方英尺①,是全美最大的会展中心。

3. retail buyers,零售商。

4. a full-fledged dinner party,大型的宴会聚餐、喜宴等。

5. trend-setting items,思想创新、样式新颖、引领时尚。

6. high-end design-forward concepts,高端设计先进的理念。

7. value-driven products,价值驱动型产品,聚焦于客户利益最大化。

8. CHESS (Chief Housewares Executive SuperSession) 指的是家居业高管会议,是业内具有策略意义的会议事件。

EXERCISES

I. Answer the following questions.

1. What is the brief history of The International Home + Housewares Show?

2. Why IHA is meaningful for housewares products industry?

3. How many expos was the Show segmented?

_____

① 1 平方英尺 ≈ 0.093 平方米。

4. What are the two critical International Housewares Association educational summits?

5. What is the prospect for The Show after 2022?

Ⅱ. Please translate the following English sentences into Chinese.

1. Exhibitors continually strained to expand their booth space, and would-be exhibitors battered at the doors.

2. The Show's primary importance was its abundance of personal contacts.

3. In 1956 no more than 649 exhibitors could fit into Navy Pier.

4. Our industry is most successful when it comes together, and it's wonderful to see the support from so many companies and brands.

5. In 1979, the West Building was pressed into service for new exhibitors with an extra day to attract buyers.

6. Retail buyers come to Chicago every year to discover new products and innovations, to meet with current suppliers and to find new opportunities for partnership.

7. Entire retail concepts are built around inventive, stylish ways to care for the home.

8. Whether a casual night in or a full-fledged dinner party, consumers cultivate their personal brands through at-home entertaining.

9. IHA will continue to work to make the 2023 Show as meaningful and productive as possible for the benefit of all attendees.

10. Technology continues to make the consumer's life easier—and with more ease comes more demand.

Ⅲ. There are 10 sentences in this section. Beneath each sentence there are four words or phrases marked A, B, C and D. Choose one word or phrase that best completes the sentence.

1. When did the International Home + Housewares Show start?? (   )

A. since 1927   B. since 1906   C. since 1949   D. since 1979

2. Conrad Hilton Hotel was previously named (   ).

A. Stevens   B. Sevens   C. Hilton Chicago   D. McCormick Place

3. How many exhibitors could fit into Navy Pier? (   )

A. more than 649   B. about 115   C. less than 649   D. more than 500

4. How could the limitation of Navy Pier be solved in 1956? (   )

A. by expanding Navy Pier   B. by providing more shows

C. by making up with on-line shows   D. by moving to McCormick Place

5. What elements make The Home and The Retailer different from others in IHA? (   )

A. new technology   B. no expense on food

C. unique and trendy designs   D. scale of show

6. Which expo offers value-driven products from China and Chinese Taiwan? (   )

A. the Clean + Contain Expo   B. the Discover + Design Expo

C. the Wired + Well Expo   D. the International Sourcing Expo

7. What is IHA's strategic and networking event for industry leaders? (   )

A. CHESS   B. CIT

C. The Lakewood Group   D. Pattern

8. What is a cooperative effort between IHA and Riemer Plus for IHA member financial executives? (   )

A. IBC Global Forum   B. The NPD Group

C. Jacob Maurer   D. The Housewares credit Group

9. Where will the Inspired Home Show 2023 take place? (   )

A. New York   B. Navy Pier

C. McCormick Place in Chicago   D. Colombia

10. According to Rob Michelson, what made it impossible for their participation in 2022? (   )

A. COVID concern   B. Schedule concern

C. budget concern   D. Economic crisis

Ⅳ. Writing

If you were one of the exhibitors who sat out the 2022 Show, write a space application for 2023 Show. You may include the following aspects.

1. To explain reasons and regrets for 2022 Show.

2. To show anticipation of participating in The Inspired Home Show 2023.

3. To apply for space application in advance.

📖 Reading 2

The National Hardware Show

The National Hardware Show (NHS) is held every year by Reed Exhibitions in Las Vegas. It is a housing after-market show that brings together manufacturers and resellers of all products used to remodel, repair, maintain and decorate the home and garden.

The Show features over 150 product categories, providing a preview of the entire home enhancement marketplace. NHS is the most comprehensive event, education and networking platform serving the hardware and home improvement industry. As the industry spearhead, it keeps a focused eye on the cultural, environmental, and technological developments shaping the future of modern living. NHS unites and guides the industry at its live events and on its interactive channels by delivering unparalleled opportunities for fostering connections, deepening insights, and building innovative and profitable strategies for growth.

The first National Hardware Show took place in New York City in 1945 and was created by Abe Rosenburg of General Tools, along with Charlie Snitow, General's Chief Legal Council. In the early 1960s, Reed Exhibitions (previously Cahners Expositions Group) acquired Snitow's trade show business. The Show moved to Chicago's new McCormick Lakeside exhibit hall in 1975. In February 2003, Reed Exhibitions announced that the National Hardware Show was moving to Las Vegas.

ABOUT THE NATIONAL HARDWARE SHOW 2023

Experience the reimagined National Hardware Show! Reconnect with colleagues and peers from all over the world at the one event that unites the industry. Home centers, independent retailers, online retailers, wholesalers, distributors, and everyone in between attend NHS to discover the newest products, innovations and trends that are shaping the future of the home improvementand DIY industry.

The National Hardware Show is located in the exciting city of Las Vegas at the Las Vegas Convention Center in the South Hall.

ONE TRIP-THREE SHOWS! In 2023 NHS will partner with NAHB International Builders' Show © (IBS) and NKBA's Kitchen & Bath Industry Show © (KBIS) at Design & Construction Week © at the Las Vegas Convention Center to offer even more buying, selling and networking opportunities to the Hardware and Home Improvement community. Use your NHS badge to enter any of the three shows.

The NAHB International Builders' Show (IBS) is the world's largest annual residential construction show and is the must-attend event for professionals in the building industry. IBS is produced by the National Association of Home Builders (NAHB) which represents 140,000+ members and is the voice for housing policies to make housing a priority.

The National Kitchen & Bath Association (NKBA) is the not-for-profit trade association that owns the Kitchen & Bath Industry Show © (KBIS), as part of Design & Construction Week © (DCW). KBIS, in conjunction with the National Kitchen & Bath Association (NKBA), is an inspiring, interactive platform that showcases the latest industry products, trends and technologies.

NHS Exhibit Hall & Featured Areas:

Explore the National Hardware Show's re-imaged floor plan featuring new and improved areas created to provide enhanced product discovery and interactive experiences. NHS debuted the launch of HABITAT in 2021, and it quickly became a must-see destination at the Show. We like to think of HABITAT as a showcase for the new ideas and technologies that are evolving how we will live at home. HABITAT at NHS can be experienced in 3 featured destinations: HABITAT Curated, HABITAT NEW, and HABITAT NEXT.

The most exciting new products from new-to-market brands and industry veterans alike come to life in HABITAT NEW, a discovery zone for the latest in Hardware and Home Improvement launches. Discover innovations by inventors whose products embody how we as an industry can deliver new solutions for everyday living and rise to meet the changes in our world. Join former Inventors Spotlight exhibitors as well as industry young guns in HABITAT NEXT.

HABITAT can also be experienced digitally year-round through curated articles showcasing products, brands, tips, and stories that embody the HABITAT Pillars. Stay tuned for the announcement of the 2023 HABITAT Pillars and catch up on previous content on the NHS Connects blog. Be sure to check out the HABITAT NEXT Stage for engaging content including Inventor Pitch Panels and educational sessions.

NHS BACKYARD is our reimagined outdoor experience, where the hardware and home improvement industry comes alive through brand activations and live demonstrations. Reaching beyond B2B through media outlets and influencers, this is an outdoor experience driven by the changing consumer landscape and the desire for meaningful brand connections.

Immerse yourself with interactive exhibit experiences you can only find in the NHS Backyard! Plus-catch engaging product demonstrations and industry influencers' 'Favorite Finds' presentations on the NHS Backyard main stage during show hours. As the day winds down, head to the NHS Backyard Beer Garden for must-attend happy hours with live music, exciting awards ceremonies, and more! Stay tuned for the 2023 NHS Backyard schedule!

If you care about high labor standards, top quality manufacturing and safer products that keep Americans working, then be sure to explore our Made in USA area for best-in-class products and services.

ENTRY REQUIREMENTS

Every participant, including attendees, exhibitors, guests and staff, are expected to comply with the health & safety requirements, including, but not limited to:

DAILY HEALTH MONITORING

All participants are asked to review and confirm they are able to comply with the Health & Safety Acknowledgement prior to entering NHS, including ensuring that they are not currently experiencing any of the symptoms of COVID-19. Anyone who is not feeling well or begins to experience symptoms at any time during NHS should contact and follow the guidance of their physician. No one is permitted entry to NHS who is not in full compliance with the health & safety requirements and the Health & Safety Acknowledgement.

Please read below for more details on our requirements, frequently asked questions, important public resources and what you can expect at NHS.

SANITIZATION AND HYGIENE

NHS has worked closely with the Las Vegas Convention Center to implement cleaning and sanitization standards that meet or exceed the recommendations of the CDC. It is the responsibility of everyone to maintain personal hygiene, including washing your hands frequently.

FOLLOW ALL SIGNAGE AND DIRECTION FROM STAFF

Please adhere to all signage and direction from all staff, including our Safety Ambassadors. We have adjusted our layout and implemented exhibitor recommendations with the wellbeing of our guests in mind. Operational adjustments may occur throughout the event.

COMPLY WITH THE CODE OF CONDUCT

Please be aware of and respect the personal boundaries of your fellow participants so that everyone can enjoy their time at NHS.

FACE COVERINGS RECOMMENDED

In accordance with public health guidance, we recommend, but do not require, that participants wear face covering while attending NHS. We encourage our participants to be up-to-date on COVID-19 vaccinations and ask that anyone who is not up-to-date wear a face covering.

HEALTH & SAFETY ACKNOWLEDGEMENT

Please ensure that you have reviewed any advisories or restrictions that may be in place for travel to the United States and for travel to Nevada. Visit https://nvhealthresponse. nv. gov/ for more information.

Please be advised that a risk of exposure to COVID-19 exists in any event or public space, including the National Hardware Show. Prior to purchasing a ticket for and attending the National Hardware Show, please ensure you have read the latest CDC Guidelines for Prevention. By attending the National Hardware Show, you acknowledge this inherent risk and, as a condition of entering the National Hardware Show, you acknowledge, understand and confirm each of the following in accordance with applicable health guidelines:

That you will abide by all National Hardware Show health-and-safety requirements;

That you are not currently experiencing any of the following symptoms of COVID-19, as identified at CDC. gov, including, but not limited to:

Fever or chills;

Cough;

Shortness of breath or difficulty breathing;

Fatigue;

Muscle or body aches;

Headache;

New loss of taste or smell;

Sore throat;

Congestion or runny nose;

Nausea or vomiting;

Diarrhea.

That prior to attending the National Hardware Show you have not been in contact with someone with confirmed or suspected COVID-19 symptoms without completing a 14-day quarantine; and That you are not under any self-quarantine orders.

Please do not enter the National Hardware Show or the Las Vegas Convention Center if you cannot confirm all of the above criteria. If at any point during your time at the National Hardware Show you do not meet all of the above criteria, you will be required to isolate and may be relocated or asked to leave the Las Vegas Convention Center at the sole discretion of event management.

Please note: The guidelines in this document are based on information currently availableregarding the behavior and characteristics of the COVID-19 virus, public health information and local or state guidelines with respect to large gatherings in any particular community. As more clarity with respect to these variables emerges, it is expected that these guidelines may need to be adjusted accordingly.

 NEW WORDS AND PHRASES

The National Hardware Show (NHS)　全美五金展

remodel　vt. 重塑

enhancement　n. 提升, 进步

foster　v. 培育, 培养, 促进

insight　n. 洞察力, 观点

unparalleled　adj. 无与伦比的, 无可比拟的

wholesaler　n. 批发商

distributor   n. 经销商, 分销商

interactive   adj. 相互作用的, 相互影响的

showcases   n. 展示柜台, 陈列柜

embody   v. 代表

compliance   n. 遵从, 服从, 顺从

physician   n. 医生, 内科医生

symptom   n. 症状, 征兆

ensure   v. 保证, 确保

hygiene   n. 卫生

vaccination   n. 疫苗注射, 疫苗接种

applicable   adj. 可以应用的, 适用的

fatigue   n. 疲劳

sore throat   嗓子痛

nausea   n. 恶心, 作呕, 反胃

vomit   v. 呕吐

Reading 3

Restart for the Automotive Aftermarket: Automechanika Frankfurt 2022

Experience innovations from international key players and learn more about new technologies and trends at the international meeting place for the manufacturing industry, repair shops and automotive trade. Like no other trade fair, it represents the entire value chain of the automotive aftermarket. Automechanika Frankfurt will be held in its familiar format as the world's leading trade fair from 13 to 17 September 2022.

Trends & Innovations

New Mobility and digitalization will permanently change the automotive aftermarket. Learn about new products and solutions for the mobility of the future at Automechanika Frankfurt. Alternative drive technologies (electric mobility, hydrogen, fuel cells, re-fuels and e-fuels, gas), connectivity and digitalization (autonomous driving, connected cars, traffic control and smart mobility) will be at the center of attention. The special show with pioneering lectures, the Automechanika Innovation Awards, product presentations, start-up pitches and research projects from universities and colleges, and a networking lounge will bring together industry players and ensure the necessary transfer of knowledge. The range of products and services is aimed at industry, trade, science and politics and also builds a bridge to the OEs. Innovations from the fields of intermodal mobility and micromobility can be found at the Future Mobility Park on the open-air site.

Workshop & Services

Digitalization, electrification, cost pressure and the need for investment (in equipment or know-how) are strongly driving the change process in the workshop business. Service concepts, safety measures and workshop equipment for the repair of electric or hybrid vehicles, access to data, diagnostics, apps, e−mobility, online portals, sustainability, recruiting and changing user be-havior are among the megatrends and challenges currently moving the automotive aftermarket. In addition to the classic range of products in the field of combustion engines, such as oil and lubri-cants, the exhibitors at Automechanika Frankfurt will be showing above all innovative solutions and new business models for the future.

Talents, Education & Training

The continuing shortage of skilled workers and the subject of further training−both are topics at next Automechanika Frankfurt. Digitalization and new technological developments pose new challenges for motor vehicle businesses every day. That is why regular further training is necessary. In cooperation with renowned partners, Automechanika offers free practical workshops that show motor vehicle and commercial vehicle professionals what is important.

11 workshops "Collison−damage repair" Body & Paint in Galleria 1:

- Free 3−hour practical workshops on 11 topics.
- All 11 workshops daily in German in the morning, in English in the afternoon (Saturdays only in the morning).
- Registration required, immediately here or on site during the fair.
- The number of participants is limited, so it's best to register now!
- For each registration you will receive a voucher for a day ticket.

Workshops:

- The workshops will be held by trainers from well−known companies in the Body & paint in-dustry−by professionals for professionals!

Special Interest Trucks, Classic Cars, Caravan

Automechanika Frankfurt is not only about cars, but also about commercial vehicles, cara-vans and classic cars. The market is lucrative for workshops, as it offers one or more additional businesses.

Around every fifth exhibiting company presents components, spare parts or tools for trucks, buses and vans at Automechanika Frankfurt. The selection of products and solutions for commercial vehicles is large. To help trade visitors quickly find relevant companies from this sector, there is the "Truck Competence" logo, which indicates what exhibitors have to offer. In addition, during the Automechanika, various activities, further training courses, entertaining impulse lectures and expert tips for everyday commercial−vehicle workshop work will be held on all days at the outdoor area F11B03.

New opportunities for workshops

Classic vehicles enjoy great popularity and represent a market volume in the billions. The busi-ness with historic vehicles is very lucrative for workshops−one or the other additional business lures here. Repair, restoration and maintenance require special expertise and craftsmanship that are not taught in today's standard trainings. Interested trade visitors will receive important first−hand infor-mation and contacts at the fair. From bodywork and vehicle technology to maintenance, repair and restoration, and from training to financing and insurance solutions, Automechanika is the only

B2B platform to present the classic car business along the entire value-added chain.

Caravanning

Campers and motorhomes are very much in vogue-the Corona pandemic has increased demand even more. An interesting market for workshops if you have the necessary knowledge. Automechanika offers special workshops on repair and maintenance for this purpose.

Automechanika Brand

The international automotive aftermarket is one of the world's most dynamic markets. Automechanika, being the leading trade fair brand, is its most important platform with 13 events worldwide. Not only is it the international meeting place for the manufacturing industry, repair shops and automotive trade, it also represents the entire automotive aftermarket value chain like no other trade fair brand.

Automechanika Navigator App

The "Automechanika Frankfurt Navigator" App is your perfect trade fair guide. With the Navigator, finding your way around the exhibition grounds is easy. The app shows you all the exhibitors and products that are most important to you and allows you to create favorites. In addition, the "Neuigkeiten" [News] feature keeps you up-to-date on the latest happenings on the exhibition grounds. The Automechanika Navigator is available free of charge for Apple devices (iPhone, iPod touch, iPad) in the App Store and for Android from Google Play Store.

Automechanika-Digital Extension

In addition to the personal Automechanika in Frankfurt, there will also be a digital offer this year. The focus of the digital extension is on pre-show and post-show preparation as well as additional networking formats for your exchange with customers, partners and journalists no matter where you are and how you participate.

Automechanika Frankfurt offers the complete event experience at the exhibition grounds from 13-17 September 2022 and, with the digital extension, also has digital networking opportunities on offer. In the digital space, you can find new contacts via AI-supported match-making or schedule appointments with your customers and business partners. In addition, every morning the Automechanika team will present the event highlights in a live stream. Selected content from the event program can be viewed as video-on-demand after the actual event.

Features at a glance

Matchmaking: network with suitable business partners.

Company profiles: Product information and company profiles of all exhibitors incl. presentation of services, highlights, products and innovations via text, image and video.

Appointments: Plan appointments for personal meetings at the fair or for video/text chats.

Downloads: Download documents and contact details.

Chat function: Use the chat function for direct exchange with other participants.

1-to-1 video call: Have face-to-face conversations in real time with suppliers and experts.

Lead manager: find all potential leads at a glance and manage them with your colleagues in real time.

Responsiveness: Responsive application for participation on the move with mobile devices.

Available all day, every day during the fair and open for exhibitors and visitors until 30 September 2022.

 NEW WORDS AND PHRASES

digitalization   n. 数字化

start-uppitch   n. 启动地, 开启活动

lounge   n. 起居室, 休息室

micromobility   n. 微观移动

diagnostics   n. 诊断学

hybrid   n. 杂交, 混合物

lubricant   n. 润滑剂

lucrative   adj. 利润丰厚的, 有利可图的

impulse   v. 推动, 促进

pandemic   n. 流行病, 传染病

restoration   n. 重建, 恢复

relevant   adj. 相关的

sustainability   n. 持续性

recruit   v. 吸收, 征募

 ENGLISH FOR WORKPLACE COMMUNICATION

Sample Dialogue: Interview with the new CEO of a software company.

Diana: With me today on Business Matters is George Jensen, the new CEO of Savannah Software. Welcome to the show, George.

George: It's great to be here.

Diana: How long have you had your feet under the desk at Savannah?

George: Well, it's been about a month now, a pretty hectic month I must say.

Diana: And how are you finding things?

George: Savannah is a very strong company, good financials, buckets of potential for future growth. There's no way I would have even considered coming to Savannah if I wasn't already sold on what a fine company it was.

Diana: But you left your previous company and went to Savannah also knowing the company had underperformed in the last few years?

George: Every company has the potential to perform better. If you didn't believe that, you wouldn't get out of bed in the morning.

Diana: So where do you see room for improvement at Savannah? What areas have you earmarked for your attention in these early days of your leadership?

George: Well, obviously I'm not going to be giving away any trade secrets here on your show, Diana, however great I think your show is! But it's no secret that Savannah has lost its focus somewhat in recent times. We need to get back to basics, which is of course high-quality, competitive software for businesses.

Diana: Where has this loss of focus been most evident in your opinion?

George: I don't think Savannah has any business trying to force itself into the home entertainment market. In time, of course, that might well change and we can reassess the software landscape at that time. For now, though, Savannah needs to return to its core business.

Diana: What is Savannah's core business? What are its flag bearing software applications? We've seen you branch out into entertainment areas such as gaming and, if I may say so, low standard educational titles.

George: Well perhaps I don't share the severity of your opinion concerning some of our recent publications, but yes we do have core applications: text to speech and speech to text applications for businesses are what put us on the map and we need to focus our attention back there because we have lost a significant amount of market share. Others have come along and beaten us at our own game, they've made improvements where we should have done so, met customer requirements where we failed. In short, we must return at Savannah to the very pinnacle of business software solutions. We can't be playing catch up anymore.

Diana: What about your prices? One of the biggest criticisms I've heard about Savannah was the disparity between quality and price.

George: We spend a huge amount on research and development. Our prices do reflect the work that goes into the development of what is very complex software. If we think we can gain market advantage by dropping our prices, that's of course something worth considering.

Diana: I read an interview you gave in Business Weekly a week or so ago. In there, you said you wanted to bring Savannah to the world, or rather the non-English speaking world.

George: I spoke before about looking for areas of potential growth, areas where we have been historically weak and I think this is certainly one of them. There's no reason why we can't produce our top language recognition software for business people in Paris, Madrid, Munich, or anywhere else for that matter. Savannah has been too inward looking, too parochial in the past. You know, the first thing I did when I got into my new office in Boston was to put a map of the world on the wall above my desk. That there wasn't one, I think, speaks volumes about some strategic mistakes that this company has made in the past.

Diana: What about the mobile revolution? As of this moment, there isn't a single Savannah app available for any mobile device and I'm thinking that so many of the products that you offer would be perfectly suitable for development as an app. Are there any plans in that direction?

George: Well, there are now! No, obviously this is something we've been thinking about for quite some time. Clearly, we should have been more proactive and had something on the market already, but we will do, there will be something by the end of the year. There's a fantastic potential for using our software on mobile devices, phones, iPads, things of that type. Our applications were designed for a fast-moving business world and their use on handheld devices is a dream come true. We think it's a match made in heaven.

Diana: George Jensen, new CEO at Savannah Software, it's been a pleasure talking to you today. We'd love you to come back and talk to us once you've been in the job for another little while longer.

George: That would be fantastic.

UNIT 7

PROTECTION OF INTELLECTUAL PROPERTY RIGHTS AT EXHIBITIONS

本章导读

知识产权对企业而言是宝贵的资产。了解知识产权保护方面的知识有助于企业通过向客户提供实用且独特的产品设计和技术，从竞争者中脱颖而出，进而建立清晰的品牌差异化战略并在未来对创新进行出售或特许经营，以增加公司收益，从而加强企业品牌化管理。

通过本章的学习，了解知识产权保护的重要性。在经济全球化快速发展的新形势下，越来越多的中国企业出国参展，国际展会已成为中国企业向世界树立形象、推广品牌、展示实力的窗口。但由于缺乏对欧盟等国家关于展会知识产权保护的了解，近年来，中国企业在参展中不断遇到有关知识产权的纠纷。为应对展会上出现的侵权行为和被侵权行为，应事先准备好合法有效的知识产权权属证明，包括商标、专利和著作权等。

知识精讲

展会知识

关于知识产权的三段案例

2004 年 2 月 15—18 日在中东迪拜举办的国际电力灯具新能源博览会上，来自福建的一家生产开关、插座类产品的企业在展览过程中赫然将某国外企业生产的同类产品放在自己的展台作样品。该企业发现后，即对侵权现场拍照并向组委会投诉，此时被投诉企业竟还称不知道此种行为侵犯了他人知识产权。最后该样品被扣押，展台亦被组委会查封。

2006 年 10 月，在巴黎举办的世界制药原料展览会（简称 CPHI）上，法国制药企业赛诺菲—安万特集团指控我国 3 家参展企业展览、交易的原料药产品侵犯了其原研药物 Rimonabant（减肥药的一种原料）的专利权。随即，这 3 家正在法国参展的 6 名医药化工贸易业人员，迅速被法国内政部扣押，并移送至当地法庭进行审理。

2010 年 2 月，在德国法兰克福国际春季消费品展览会上，河北一家出口玻璃制品的企业被德国海关指证产品侵权，要求该企业将展品撤下展台，并写出保证不再展出侵权商品。同年 8 月，该公司在法兰克福参加秋季消费品展览会，又一次展出了涉及侵权的玻璃制品，再次被海关查到。该企业遭到现场展品被没收、罚款 500 欧元的处罚。

思考并回答以下问题：

1. 以上案例集中反映了国际展览会中的什么问题？

2. 通过这些案例，分析我国外贸企业做到哪些方面才能提高在国际市场上的竞争力？

 **Reading 1**

UFI Recommendations for the Protection of
Intellectual Property Rights at Exhibitions

An exhibition, as the marketplace of an industry sector, is the perfect location for product and service counterfeiters to undertake illegal practices. However, exhibitions also contribute in the fight against these practices as they represent an easy way for manufacturers and service providers to identify counterfeited products and services, and potential threats. Exhibitions provide excellent opportunities to obtain information on competitors and to discover the existence of new products and services, hence pinpointing potential IPR infringements at the initial stage prior to large scale manufacturing and commercialization.

One of the missions of UFI, The Global Association of the Exhibition Industry, is to help its members and the exhibition industry in defending business interests, whilst promoting exhibitions as the most powerful marketing, sales and communications tool. In this respect, UFI has drawn up recommendations to be used by any exhibition organizer to assist their clients-the exhibitors and the visitors-in the protection of their Intellectual Property Rights and in the defense of these rights if infringed or endangered. This document has been elaborated to provide guidance on the necessary measures to be taken for effective IPR protection and enforcement during trade fairs. The scope is worldwide and hence covers generalities which are applicable in most countries. More in-depth or more rigorous actions depending on a specific country may be required.

Piracy and counterfeiting are similar terms employed to describe the illegal reproduction or imitation of products/services that infringe intellectual property rights. Intellectual Property (IP) can be divided into two categories:

• Industrial Property, which includes trademarks, patents, utility models and designs;

• Copyright and neighboring rights, which includes literary and artistic works such as novels, poems and plays, films, musical works, artistic works such as drawings, paintings, photographs and sculptures, and architectural designs.

Trademarks (or its commercial equivalent brands)

A trademark identifies and distinguishes the products or services of a company from those of its competitors. Several types of trademarks exist: e. g. word marks, device marks (logos), three dimensional marks, color marks or combinations thereof. The protection of a trademark provides the holder with an exclusive right of use and the potential to defend against the use and registration of identical or similar trademarks which could cause confusion in the public eye. The protection of a trademark is generally for a 10-year period and can be renewed. The protection of a trademark is effective on a territorial level (i. e. the exclusive right exists only in the countries in which the trademark has been registered). A single protection is possible for the entire European Union (currently 27 member states).

Patents

Patents cover inventions and protect the (technical) characteristics of a product or a process. The patent owner can therefore prevent a third party from exploiting the invention. He may use the invention provided the product or process contains no features covered by another patent. Otherwise, the parties involved should consider cross-licensing their rights. The duration of a patent is generally 20 years and is subject to fee payment (usually annual). It is also possible to renew the patent. The protection offered by a patent is effective on a territorial level. A single procedure is available in Europe (currently allowing protection for up to 37 counties) through one examination in one language.

The requirements for the granting of a patent are:

Because of the novelty, inventiveness or non-obviousness, it is crucial that a susceptible of industrial application or usefulness. The requirements for the granting of a patent are in most countries, there must be a non-obvious technical contribution to the state-of-the-art. The state-of-the-art is formed by everything already known to the public before the filing of a patent application (even disclosures by the patent applicant themselves).

Utility Models, Petty Patents and Other Types of Patents

The utility model and petty patent are also rights which genuinely protect an invention. The requirements for utility models and petty patents are usually the same as those for patents. In most cases, there is no detailed examination as to the validity of such rights. For this reason, a utility model can be obtained quickly, easily and cost-effectively. The duration is shorter than for a regular patent, and the protection may be narrower in some countries. Other types of "patents" exist in some countries, such as plant patents or plant variety protection to protect a new variety of plants.

Designs

A design protects the appearance of the whole or a part of an industrial or handicraft product determined, in particular, by the lines, contours, colors, shape, texture and/or materials of the product itself and/or its ornamentation, provided this appearance was new and had a so-called individual character at the time of filing the application. A design can be two-dimensional or three-dimensional. The duration of protection varies from country to country. In the European Union for example, a registered design is valid during 5 years, and can be renewed for up to a maximum of 25 years. The protection of a design is effective on a territorial level. A single protection is possible for the whole European Union (currently 27 member states).

Copyright

A copyright protects original creations, such as literary and artistic material, music, films, sound recordings and broadcasts, including software and multimedia. In the US, it is recommended to file a copyright. In most other countries, it is not possible to do so. Like any property, IP can be rented (licensed) or sold (assigned). Due to the complexity of the IP registration procedure, the use of the services of IPR professionals is highly recommended. In some countries, unregistered designs and trademarks (but not patents) may be protected, but usually for a much shorter and reduced protection. Claims arising from the law against unfair competition may apply even if the copied product is not protected by the above-mentioned property rights, but only under special circumstances, i. e. , when the imitation is slavish.

Trademarks, patents and designs registered in one country are not protected in other countries. For example, US trademarks, patents and design rights alone do not provide protection in the countries of the European Union and vice-versa. In the same way, the IP acquisition or registration procedure may differ from one country to another. In particular, IP acquisition in the USA differs substantially from most other countries.

Whilst acquisition and protection of IPR may not be similar from one country to another, international agreements to harmonize some of the main aspects of the IP laws have been adhered to by most countries. The most important agreement in this regard is TRIPS (Agreement on Trade-Related Aspects of Intellectual Property Rights) enacted by the World Trade Organization.

The owner of an IP right can enforce that right through different actions-warning/request to discontinue use letters to infringers; customs actions; court actions; dispute-settling by mediation and arbitration. The collection of proof (in that case and in many countries, a "descriptive seizure" procedure or ("Anton Piller" order) enables the IP right owner to obtain the decision through the Courts to send a neutral expert to the premises (offices, exhibitions) of an alleged infringer in order to describe the alleged infringement and seize evidence.

In conformity with the laws of many countries, the exhibition organizers are usually not legally authorized to take action against infringements. Only the exhibitors (IP right owners), or their lawyers, can take effective action against infringers (pirates or counterfeiters).

An effective method of preventing the exhibiting of plagiarized copies before a fair opens is border seizure by the customs authorities. Customs offices check merchandise that is being imported or exported or is in transit to see whether it contains products that infringe protected rights. Their work is not limited to inspections at border stations but also includes, as part of industrial property protection, monitoring and inspection at border customs offices, inland customs offices and free ports or by mobile control units. On receipt of an application showing the right of a brand manufacturer, customs authorities can stop suspicious consignments, examine them, take samples, destroy counterfeits and supply information to the brand manufacturer.

During the trade fair, an exhibitor can take the following measures with the help of a lawyer:

● Declaration to cease and desist: The "copier" signs an undertaking that he will no longer offer the copied products for sale and will pay a fine in the event of a further infringement.

● Preliminary Injunction: The holder of the patent or design right obtains a temporary court injunction prohibiting the "copier" from selling and exhibiting certain products.

UFI Recommendations

Exhibition organizers should ensure an equitable business environment during trade shows by informing, protecting and assisting their exhibitors in acting against brand and product piracy.

Before the event, organizers should provide exhibitors with information on IPR protection via a specific brochure to be provided with the registration/participation forms, on the organizer's website, in the exhibitors' manual or in the trade show's "General Terms and Conditions". This information should contain general advice for exhibitors and include the following recommendations:

● that exhibitors protect and register trademarks, patents or designs before the trade show starts, to obtain a valid right (an exhibition destroys novelty) and hence use all forms of legal protection, both in general and during the event.

● the use of a specialized patent and trademark lawyer regarding registration alternatives, requirements, procedures and maintenance.

● exhibitors should bring to the trade fair all original documents or certified copies of their patent or trademark rights, so that a possible infringement can be established during the event. Any verdict already obtained against an exhibiting pirate should also be included.

● exhibitors should be encouraged to indicate that their products or services are protected by IP rights, where applicable.

● if an exhibitor believes that another exhibitor will infringe their rights then they should make the appropriate application to the customs authorities (when applicable), who can then stop suspicious consignments, investigate them, take samples, and destroy copies. This should of course take place before the exhibition.

In addition, the organizer should also provide both before and during the trade fair: the contact details of the person responsible for IPR issues within the organizing company, the contact details of local/national IPR organizations, customs authorities and patent and trademark lawyers willing to represent exhibitors who wish to pursue legal action against an alleged infringer. This may include the possible subsequent identification of counterfeit products during the trade fair.

Organizers should be able to provide a neutral arbitration, arbitrator, or judge to help determine if there is a violation or to resolve IPR disputes during the trade fair, and should provide interpreters to facilitate communication in the case of disputes with foreign exhibitors. When appropriate and if possible, organizers should provide an on-site office, a special stand or a point of contact, to deal with any IPR requests or complaints for the entire duration of the trade fair.

(This article is edited according to Internet information)

 ### NEW WORDS AND PHRASES

counterfeiter   n. 伪造者

IPR(Intellectual Property Rights)   知识产权

infringement   n. 侵权;违反

utility model and design   实用新型和设计

piracy   n. 盗版行为

cross-licensing   n. 交换许可

susceptible   adj. 易受影响的;可以接受或允许的

state-of-the-art   adj. 使用最先进技术的

copyright   n. 版权,著作权

acquisition   n. 采集

harmonize   vt. 使和谐;为(旋律)配和声   vi. 和谐;以和声演奏或歌唱

infringer　n.（法）侵权人

arbitration　n. 仲裁；公断

seizure　n. 没收；夺取；捕捉

plagiarize　v. 剽窃，抄袭

 NOTES

1. TRIPS 是 Agreement on Trade-Related Aspects of Intellectual Property Rights 的缩写，即《与贸易有关的知识产权协议》。

2. "Anton Piller" order，"安东·皮勒"令。其是一项扣押令，旨在防止被告破坏证据，实际上是一项民事搜查令。

3. Preliminary Injunction，预先禁令。也叫"中间禁令"或"临时禁令"，指起诉后、判决前由法院签发的禁令；禁止被告实施或继续某项行为。

 EXERCISES

Ⅰ. Answer the following questions.

1. What are the functions of UFI?

2. Could you explain the classification of intellectual property?

3. What is a trade mark?

4. What is a patent?

5. What are the requirements for granting patents?

6. What actions can owners of intellectual property enforce their rights?

Ⅱ. Please translate the following English sentences into Chinese.

1. Exhibitions provide excellent opportunities to obtain information on competitors and to discover the existence of new products and services, hence pinpointing potential IPR infringements at the initial stage prior to large scale manufacturing and commercialization.

2. One of the missions of UFI, The Global Association of the Exhibition Industry, is to help its members and the exhibition industry in defending business interests, whilst promoting exhibitions as the most powerful marketing, sales and communications tool.

3. The protection of a trademark is generally for a 10-year period and can be renewed.

4. The protection of a trademark is effective on a territorial level (i. e. , the exclusive right exists only in the countries in which the trademark has been registered).

5. A single protection is possible for the entire European Union (currently 27 member states).

6. The duration of a patent is generally 20 years and is subject to fee payment (usually annual). It is also possible to renew the patent.

7. The protection offered by a patent is effective on a territorial level. A single procedure is available in Europe (currently allowing protection for up to 37 counties) through one examination in one language.

8. The exhibitors protect and register trademarks, patents or designs before the trade show

starts, to obtain a valid right (an exhibition destroys novelty) and hence use all forms of legal protection, both in general and during the event.

9. Exhibitors should be encouraged to indicate that their products or services are protected by IP rights, where applicable.

10. Organizers should be able to provide a neutral arbitration, arbitrator, or judge to help determine if there is a violation or to resolve IPR disputes during the trade fair, and should provide interpreters to facilitate communication in the case of disputes with foreign exhibitors.

Ⅲ. There are 10 sentences in this section. Beneath each sentence there are four words or phrases marked A, B, C and D. Choose one word or phrase that best completes the sentence.

1. Which of the following is the best translation of the phrase "non-obviousness"? (　　)

A. 不明白　　　　　B. 不清楚　　　　　C. 不显著　　　　　D. 非显而易见

2. The phrase "industrial application" most probably means (　　).

A. 工业应用　　　　B. 工业申请　　　　C. 工业施用　　　　D. 工业运用

3. The holder of the patent or design right obtains a temporary court injunction prohibiting the "copier" from selling and exhibiting certain products. "copier" probably means (　　).

A. duplicate　　　　B. model　　　　　C. reproduction　　　　D. infringing product

4. The phrase "product piracy" is also called (　　).

A. product hijacking　　　　　　　　B. an illegal copy product

C. product theft　　　　　　　　　　D. product duplicate

5. The phrase "a regular patent" is also called (　　).

A. conventional patent　　　　　　　B. specific patent

C. frequent patent　　　　　　　　　D. elderly patent

6. The phrase "descriptive seizure" probably means (　　).

A. poor health

B. in bad condition

C. evidence of preliminary suspicion of infringement of intellectual property rights

D. evidence of infringement

7. the phrase "UFI Recommendation" means (　　).

A. UFI Suggestion　　B. UFI Order　　　C. UFI Permit　　　D. UFI Instruction

8. Which of the following is not covered by copyright protection? (　　)

A. original software　　　　　　　　B. original multimedia

C. imitation film　　　　　　　　　　D. original recording

9. Which of the following is not included by copyright? (　　)

A. paintings　　　　B. trademarks　　　C. photographs　　　D. sculptures

10. What of the following is the closest in meaning with the phrase "Anton Piller" order?
(　　)

A. temporary injunction　　　　　　　B. interlocutory injunction

C. civil search warrant　　　　　　　　D. trade order

Ⅳ. Writing

Please write a composition about "How do enterprises strengthen intellectual property protection in international exhibitions" with the following particulars(Not less than 200 words) :

Some problems

1. Most enterprises have weak awareness of intellectual property protection.

2. Many enterprises have not established special intellectual property protection institutions.

3. Many enterprises also didn't hire professional talents for consultation.

Effective strategies

1. Improve laws and regulations on intellectual property protection.

2. Attach importance to the training of intellectual property related talents.

 **Reading 2**

Messe Frankfurt Against Copying-Our Initiative for Protecting Your Rights

Protecting intellectual property is the basis for fair competition. Since 2006, Messe Frankfurt against Copying has worked to ensure that your innovations are well protected at our trade fairs.

Be sure to make full use of the rights available to you in Germany and to do so in good time. We would be happy to assist you in any way we can.

What do you definitely need to know? Industrial property protection is your basis for initiating legal proceedings against infringements of intellectual property rights. Only the person whose intellectual property rights have been infringed can assert claims arising from the relevant legal provisions.

Here you can find the most important steps for protecting your products—both before and during your trade fair activities. Better safe than sorry …

Before the trade fair:

How and where can you find out about registering intellectual property rights?

Your first step should be to contact the German Patent and Trade Mark Office (GPMA) , the central institution for protecting intellectual property in Germany. It is here that intellectual property rights are issued and administered. The German Patent and Trade Mark Office is a partner in a network of national, European and international intellectual property rights systems. Further competent partners include the regionally based patent information centers. These provide practical assistance on all questions relating to protecting intellectual property and work closely together with the German Patent and Trade Mark Office.

Please note that the nature of the intellectual property rights determines the jurisdiction of the registration authority. Accordingly, you should inform yourself about your specific situation without delay. The names of registration authorities outside Germany and details on individual intellectual property rights can be found on the GPMA cooperations website.

Have you already registered intellectual property rights for your products?

Please check the time periods and geographic areas for which your intellectual property rights are valid. Intellectual property rights must be registered for the country in which the trade fair in question is taking place. You should always carry original documents, certified copies or the registration numbers with you for the purposes of verifying registered intellectual property rights. If any final verdicts have already been returned against the imitator in question, these are also essential as proof.

Do you expect infringements by third parties at our trade fair?

You should attempt to resolve infringement issues prior to the trade fair. If necessary, please contact the potential infringer after seeking counsel on this from your legal advisor.

What else can you do beforehand?

One effective way of preventing counterfeit and illegally manufactured products from being presented at trade fairs is to file an application for action by customs authorities ("border seizures"). Goods from non−EU countries entering the country for exhibition at trade fairs can be inspected for possible infringement of intellectual property rights by customs officials before the trade fair takes place.

And during the fair:

What if an infringement of intellectual property rights comes to my attention during the event?

You should take immediate action. Contact the emergency legal service or your lawyer as soon as possible. You can make use of an initial consultation on location through our emergency legal service. During the trade fair opening times, you can reach the lawyers at the following hotline number: +49 69 7575−1212. Needless to say, you can also seek advice from your own lawyer. Together with your chosen legal advisor, you can decide on which legal steps you would like to take.

Which are available legal steps for enforcing your intellectual property rights?

Civil law measures: Cease−and−desist declaration:

The potential imitator agrees to cease the infringing behavior (exhibiting, promoting or in some cases selling copied products). In the event of a further infringement, a contractual penalty is to be paid. You can instruct your legal advisor to formulate a cease−and−desist declaration.

Preliminary injunction:

A temporary court order prohibits the actions of the potential infringer (exhibiting, promoting or in some cases selling copied products). A preliminary injunction can be obtained from the court by your legal advisor.

Criminal law measures:

Another possibility for combating infringements of intellectual property rights is to file a criminal charge with the police. The police are authorized to take action on condition that there is a suspected infringement of intellectual property rights on the part of an exhibitor from the European Union. The police station is found in the Operation Security Center in Hall 4.0. Southwest and can be contacted by telephone at +49 (0) 69 75 75−65 55.

In the event of infringements of intellectual property rights on the part of exhibitors from non−EU countries, customs officials can take action by carrying out inspections. Please note that customs officials carry out their inspections independently and at selected events. Please find out in good time before the trade fair begins about the procedure for applying to take part in a customs inspection. To do so, please contact the Main Customs Office in Darmstadt.

What steps can you take to prevent people taking unauthorized photographs of your products?

Please contact Messe Frankfurt's security service at +49 69 75 75－33 33. Before and during a trade fair, stickers indicating that photography is prohibited can be obtained for your stand fromthe trade fair management team.

(This article is edited according to Internet information)

 NEW WORDS AND PHRASES

initiating　v. 开始(initiate 的现在分词); 传授; 发起
assert　v. 坚称, 断言; 维护, 坚持; 坚持主张
jurisdiction　n. 司法权; 管辖权; 管辖范围; 权限
verdict　n. (陪审团的) 裁决; 裁定; (经过试验、检验或体验发表的) 决定; 意见
legal proceeding　法定程序, 法律诉讼
infringements of intellectual property rights　侵犯知识产权的行为
legal provision　(法)法律规定
registration authority　注册机构
geographic area　地理区域
contractual penalty　合同罚款

 **Reading 3**

How Exhibitors Can Protect Their Intellectual Property at Trade Shows

It goes without saying that trade shows can be frontiers of innovation and creativity, an incredible forum for viewing what industries are developing as the next best products or ideas. In only a matter of days, a leading trade show can display hundreds or even thousands of products, services and ideas in front of a global audience.

Yet all this added exposure on the show floor can also serve as a catalyst for intellectual property disputes, including over trademarks, copyrights, trade secrets and patents on developing inventions and designs.

So, what can exhibiting companies do to protect their intellectual property and ideas? To address this pressing industry issue, TSNN consulted the experts, in this case, Dan Cleveland, intellectual property attorney at Fennemore Craig, to get his legal expertise on the importance of IP protection and how exhibitors can take action against IP infringement.

TSNN: Has your law firm been involved with any IP disputes that had their origins at trade shows or other events?

Dan Cleveland: I am relatively new to Fennemore Craig, although I have been practicing law for more than 20 years. In the past year, I have been involved in two IP disputes with trade show origins. They are still ongoing or have recently settled, so I cannot provide specific comments on those particular disputes. However, I am able to say that I have had involvement or have "closely-witnessed" seizure of falsely designated NCAA jerseys and T-shirts, computer components and medical devices with origins at trade shows.

TSNN: In what specific ways can intellectual property be at risk at trade shows?

DC: It can be particularly galling for an exhibitor to sit at a trade show and watch a competitor gain commercial advantage through intellectual property infringement. There are different types of intellectual property and we primarily see the protections afforded by trade secrets, trademarks and patents being at risk at trade shows.

Trademarks protect the goodwill of your business. The purchase price for most companies far exceeds the value of physical assets that the company owns. The difference between this value and the purchase price is goodwill. In a trademark sense, this is the established expectation that customers have about the kind and quality of goods or services which the company provides. It is unfair if a competitor uses a trademark that gives consumers a false impression of origin for goods or services and so gains commercial advantage by misdirecting these consumers. This not only costs money by hitchhiking on your good name but also can damage your reputation by providing goods or services of inferior quality.

Patents are another type of intellectual property that can also be enforced at trade shows but this is more difficult to do. Patents are frequently used to distinguish products in the marketplace because a patent owner can exclude others from practicing what is claimed in the patent. It is commonly believed that having a patent in effect is a way to achieve higher margins on product sales. In "core technology" cases, patents may even entirely exclude competitors from direct competition with a product line, which is why you often see extremely high manufacturing margins in the case of on-patent pharmaceuticals.

If left unchecked at the trade show, patent infringement denies you the benefits of being an innovator as customers go elsewhere because you have higher prices. This also discourages innovation. Design patents protect the ornamental design of a product and this can be easy to ascertain. Utility patents are frequently more complex and may require more study before a case for infringement can be made. Even so, there are also simple utility patents and complexity is less of a problem if one has a bit of time in advance to study a competitor's product.

Utility patents protect ideas or concepts and have claims that are directed towards combinations of limitations. To illustrate this, we can describe a breakfast that includes a limitation of bacon, eggs, toast and a sleeping pill. The standard breakfast of bacon, eggs and toast is un-patentable because it has been known forever. On the other hand, the Patent Office might be persuaded that adding a sleeping pill to a breakfast is patentable because most people would not want to go to sleep right after breakfast.

A trade secret is something that has inherent value due to the fact that it is not generally known to the public. A common concern at trade shows is the tradeoff of wanting to make sales by showing customers advanced technology, versus the risk of exposing key aspects of technology to competitors who may also visit your display. If you have these issues then you need to train the sales force manning your display not to over-disclose into the area of trade secrets.

TSNN: What are some key things trade show exhibitors should be aware of when exhibiting their products and services?

DC: Comparative statements and advertising are a fine thing to do, provided the comparison is strictly true and not misleading. Exhibitors should not be afraid to make these kinds of comparisons and may even call out competing products by their trademarked names. In doing so, however, they should be should be careful not to overstate their case into the realm of misleading prospective customers.

Also, if exhibitors know that they are functioning in an illegal manner or even in a gray area, they should be aware that it is possible for the offended party to get a court order and bring in a federal marshal to shut this down right on the trade show floor. This can even happen without having first brought them into court, but a caveat exists that the procedure for getting the court order requires the posting of a bond and the "infringer" must be reimbursed if there is a wrongful seizure. On the other hand, the same procedure is available as a remedy if an exhibitor is being harmed by knock-off products.

TSNN: What should exhibitors keep in mind when attempting to protect and enforce their IP rights?

DC: While the remedy is available, it does take some advance time to work through the process. Never say never, but since most trade shows last only a few days it is difficult to witness an instance of intellectual property infringement at a trade show as a matter of first instance and take action in time for a seizure order to make a difference. Even so, it is possible to use the awareness of infringement and soon after take action.

As you might imagine, a seizure order is a form of extraordinary relief. You have to meet strict requirements pursuant to statute and persuade a judge to approve an order that does not overreach in view of the concerns you are stating. One of the things usually placed at issue is whether irreparable harm will occur if the order does not issue. A lack of irreparable harm may be presumed from undue delay, so if you are the offended party this is not the time to sit on your rights and nurse a grudge.

(This article is edited according to Internet information)

 NEW WORDS AND PHRASES

attorney　n. 律师; 代理人

jersey　n. 毛织运动衫, 毛线衫

galling　adj. 使烦恼的, 难堪的, 使焦躁的

ornamental　adj. 装饰的, 装饰用的

pharmaceutical　adj. 制药的, 配药的　n. 药物

patentable　adj. 可给予专利权的; 可取得专利证的; 可给予专利证的; 可取得专利权的

over-disclose　v. 过度披露

tradeoff　n. (公平) 交易, 折中, 权衡

knock-off　n. 假冒商品; 敲去

extraordinary　adj. 非凡的; 非常奇特的; 特别的

irreparable   adj. 不能修复的, 不可弥补的; 不能挽回的; 无可补救

grudge   n. 不满; 怨恨; 恶意; 妒忌

ENGLISH FOR WORKPLACE COMMUNICATION

Sample Dialogue: An Introduction to Intellectual Property Right.

Situation: The following conversation is between Bob Legvold from Columbia University (B), astudent named Lily (A), they are talking about Intellectual Property Rights.

A: Hello, professor, I am here for a consultation regarding *intellectual property rights*.

B: Yes, please.

A: What is the difference between copyrights and trademarks?

B: Copyrights are used to protect authored or created works such as art, music, film, and writing. Trademarks apply to single words, phrases, and logos used with merchandising of a good. Some situations may require protection with both a copyright and a trademark.

A: How to apply for copyright and trademark?

B: Applying for a copyright or trademark involves specific processes. To apply for a copyright, the applicant pays a filing fee and a short registration period ensues. The U. S. Copyright Office will perform a review of the forms and application before approving the copyright. Registering for a trademark is more expensive and lengthy process because the federal trademark office conducts an exhaustive review to make sure that the trademarked content is unique and not like any other trademarked content.

A: Sounds great! While, how to protect intellectual property?

B: I think make time to get smart on intellectual property. Educate yourself and team on the basics of trademarks, copyrights, patents, and trade secrets. Investing a day or two early on will save headaches later. What's more, work with an attorney who specializes in intellectual property and ask for a fixed rate to file.

A: Thank you very much!

B: It's my pleasure.

UNIT 8

BUDGETS IN TRADE FAIR

本章导读

　　展览会的预算涉及很多方面，如光地、装修、物流、人员等。许多参展商将其与展会管理的互动局限于后勤事务，如购买展位、联系装修等。但是，如果与展会主办方管理层分享目标并进行合作，则可以协商额外费用，例如赞助、每日展会广告、演讲机会、参加小组会议，就有机会降低成本或得到一些免费的服务，如更多的参展商证件等。

　　随着全球疫情以及经济的不景气，很多展览会面临着部分展商会缩小展位规模的局面，因为参展商会重新评估展出效果，除重新考虑展品的组成外，还会为更多的社交距离腾出空间。一些公司会缩减市场营销预算，减少展会上的工作人员。在缩小了面积的展位上如何显示出较大的空间，是摆在展位设计及搭建商面前的一个挑战。

　　对于常年参展的公司来说，一套展具可以重复使用，但每次展会都存在展具磨损，时间一长就会显旧。翻新旧展具也是一个技术活儿，如何用最小的预算获得好的效果，在本单元第二篇文章中进行了详细的描述。

知识精讲

展会知识

集结慕尼黑，全球工程机械行业盛会即将启幕

　　德国慕尼黑国际工程机械宝马展会 BAUMA 是全球规模最大、最专业的建筑、矿山及工程机械展会，始创至今已有 50 多年历史，其展品范围全面包括了全世界各类建筑机械、设备及工程用车和矿山机械，它不仅是国际建筑业界的商务贸易中心，同时也是世界各地建筑业者汇聚交流、获取信息、拓展联络的重要平台。

　　2019 年 4 月 8 日至 14 日，宝马展在德国慕尼黑盛大举办。三年一届的工程机械行业盛事，每一位热爱工程机械的人都不会错过，据说到了 BAUMA 展现场就可以感受行业的心跳，全球 60 多个国家和地区的 3 500 多家企业惊艳亮相，来自 200 多个国家和地区的超过 600 000 名专业观众到场参观。

　　这场高度国际化的盛会，集结了工程机械行业的前沿新技术，呈现令人惊呼的科技水平，心动的你计划飞往慕尼黑了吗？这个世界大展的面积竟然又扩大了，室内展馆从 16 个

增加到 18 个，总体展出面积突破 614 000 平方米，意味着更优化的展示布局与商业空间。

BAUMA 2019 释放三大行业趋势，当前工业化的三大趋势是可持续性、数字化和效率。

首先是可持续性，可持续性包括节能减排、清洁灰尘、电动替代等。

其次是数字化，很多人将数字化工厂称为"未来工厂"，网络化建筑工地、人工智能、数字模型、云空间、大数据分析、无人驾驶和机器人，这些构成我们脑海中的未来工厂雏形。

最后是效率，聚焦本次展会的建筑材料科研成果——碳混凝土。此外，目前业内已出现了无人机控制机器人施工车辆等智能化手段，进一步将人工智能运用到建筑施工领域。

另外，国际大牌蓄势待发。这个集结高品质企业的盛会，自带超强磁场，工程机械行业巨擘悉数到场。你所关注的国际展商，他们将带来怎样的精彩展示，我们拭目以待！

卡特彼勒："重新定义"是卡特彼勒本届展会的主题，展区将展出 64 台机器，其中 20 台机器是新推出的产品，展示区将强调设备用户与卡特彼勒合作后获得的全新高效率和营利能力。

维特根：维特根集团代表着智能协同与创新。约 120 款展出机型（其中包含多款全球首秀机型）及 13 000 平方米的展位面积，将展示能够理想兼容的产品解决方案以及应用工艺，使客户能够直面筑养路施工中的挑战，经济高效地施工，并获得出色的施工效果。

沃尔沃：沃尔沃此次参展的主题为"筑造未来"。室内展台面积达 2 293 平方米，露天展台面积为 5 870 平方米。展会期间，沃尔沃通过不间断的互动表演，将其着眼于未来的建筑而打造的全系产品及整体服务解决方案呈现给全球用户，并承诺小型机械将向电气化转型。

利勃海尔：利勃海尔以 Together. Now & Tomorrow 为主题，在室外 14 000 平方米的空间展示 60 余种产品，包括塔式起重机、履带式起重机、原料处理、土方工程、采矿及零部件等产品的创新与开发。

安百拓：安百拓将展示一系列创新和高效的产品，包括全新技术的液压剪、新的地下采矿、数字解决方案等。

小松欧洲：展出面积达室内 4 000 平方米、室外演示区域 2 600 平方米，展出一系列创新技术水平的 30 台全新的工程机械设备。小松的数字化创新和环境技术，将在 BAUMA 上首映。

安迈集团：安迈集团将在这次展会展出 150 多台设备，其中不乏新品。ABC ValueTec 沥青搅拌站是为提高生产效率而设计制造，最终目的是通过降低成本来提高收益。与此同时，安迈集团还将在 BAUMA 2019 期间庆祝成立 150 周年。

特雷克斯：超过 5 000 平方米的令人印象深刻的展台，将展出 40 多台机器，展出的产品代表着对研发的创新和投资。例如远程信息处理解决方案，将展示它们如何为您提供优势从而提供真正的竞争优势。

因特威：快速接头及高压球阀行业西班牙老牌企业，为世界各地工业、农业、船舶业、石油装备、航空航天、军工科技提供液压快速接头服务和解决方案。

思考并回答以下问题：

1. BAUMA 是展会的英文名字，根据发音中国人习惯称之为宝马展。查询相关资料了解宝马展的盛况。

2. 本报道中提到多个世界知名工程机械的品牌，请查找这些公司的英文名字，了解这些公司的概况，并分享给大家。

3. 根据宝马展的展品范围，你觉得中国哪些大型企业可以参加该展览会？查询展会官网资料，以我国某企业申请300平方米光地为例，做一个参展预算。

4. 除德国宝马展外，世界上还有哪些工程机械展？比如美国、法国等国家。收集这些展会的资料，并分享给大家。

 **Reading 1**

Let's Make a Deal

One of the benefits of doing business in a slow economy is that show management and industry suppliers really want and need your business. As a result, they're typically more willing to work with you to negotiate prices, services, discounts, and promotional opportunities.

But don't get greedy. Negotiation is not trickery; it's coming to a mutually agreeable solution that will benefit both parties. Whether you're negotiating with new vendors, or revisiting terms with existing ones, negotiation is about developing relationships based on ethics, truth, and honesty, not getting adversarial, emotional, or angry.

Before negotiating, do your homework. Familiarize yourself with the state of the market. Know what vendors are charging and what items are commonly negotiable. Next, identify which issues are most important to you. Are you looking for a solution to a problem, a lower price, better service? How flexible can you be with what you're negotiating for? Compile a list of the needs that you have to have and wants that would be nice, but aren't necessarily deal breakers.

Finally, figure out your acceptable bottom line ahead of time, and be ready with a plan B. What's your fallback position if you can't reach an acceptable deal with your vendor? Do you have a back-up supplier in the wings that you can go into negotiations with? Don't burn your bridges unless you have a solid plan B.

The following list represents a variety of the people and suppliers with whom you can negotiate throughout the trade show management process. Use this information as a guideline to help you prepare for your negotiations—and proactively identify what is and what is not negotiable to begin with—and you're far more likely to get what you're asking for.

Show Management

Many exhibitors limit their interactions with show management to logistical matters such as purchasing booth space. But if you share your objectives and work with show management, you can often negotiate for extras, such as sponsorships, advertising in the show daily, speaking opportunities or participation in panels at the conference, and free or reduced-cost exhibitor badges.

And although booth-space pricing isn't usually negotiable, I have negotiated multi-show discounts for booth-space contracts. I have also negotiated a change of due date on a booth-space rental deposit for a client. So, if cash flow is an issue for your company, it doesn't hurt to ask for an exemption.

Exhibit Houses

There are many opportunities to negotiate with your exhibit house. For example, many exhibit houses charge storage fees based on the amount of cubic feet from the floor of your storage space to the ceiling, regardless of whether your stored items actually occupy all of that available space. You can often get them to charge you for actual space instead, which can significantly lower this line-item expense. Furthermore, you can negotiate for a flat rate on fees that are usually charged hourly, such as warehousing or preparing your exhibit for a show. Turnkey fees may also be negotiable.

I was able to eliminate the exhibit-disposal fees quoted by one exhibit house by getting it to recycle the aluminum from my client's booth. The recycling income offset the landfill fees for the rest of the structure.

Also remember that you're not the only one trying to get a deal. Your exhibit house also negotiates with its vendors, such as exhibit shippers and exhibitor-appointed contractors (EACs). But it may not pass along these discounts to you and might still be marking them up by 25 percent to 30 percent or more, effectively doubling the cost of the services it orders for you. So, ask it to pass along the discounts it gets from vendors when negotiating on your behalf.

Transportation Carriers

Loyalty can go a long way when negotiating with your transportation carrier. You can generally request discounts on airfreight costs, for instance, based on volume or a record of timely payment.

If your carrier will be picking up freight for multiple exhibitors at your show, it may be charging each of those exhibitors for wait time in the marshalling yard. Ask if you can split the charges, get a flat rate on wait time, or even waive fees for wait time. Also ask your carrier to waive accessorial charges such as fuel surcharges for local moves.

Installation and Dismantle Contractors

I recently rented an exhibit from a general services contractor (GSC) that wanted me to pay for a guaranteed number of I&D hours that I thought was about three times what it would actually take to put up and take down my exhibit. After a bit of negotiation, I got the contractor to agree that I would pay for the actual number of hours it took instead.

If you are using an EAC, you can ask it to reduce its city rate to match that of the GSC's labor. You can also ask your EAC to waive the supervisory fee, its minimum-hours requirement, and reduce the charges for supplies and materials (many of which you can provide at a lower cost).

Other Vendors

You can negotiate with almost any of the other vendors you use for trade show services, such as audiovisual contractors, talent companies, promotional-products suppliers, and furniture-rental companies. For example, I recently negotiated with a show's official AV contractor to rent a larger television screen for the cost of a smaller screen, since I knew it had extra-large screens available. I've also negotiated away the fees for speakers, keyboards, stands, wall brackets for monitors, and draping.

With talent companies, I've been able to negotiate away the 10-percent agency fee they often charge in addition to the stated daily rate for talent. And in the case of rental furniture, I've negotiated multi-city contracts at discounted rates.

Talk with your show electrician about how to reduce your electrical cost based on your minimum requirements. Ask about providing your own cabling/surge protectors, request that the electrician lay your electrical on straight-time hours, and find out if it will save you money to build a power distribution unit (electrical box) to break down large quantities of power to cheaper, smaller outlets.

If you make reasonable requests and build strong partnerships with your suppliers, they will work with you to meet your objectives and your budget. Whether you're working with show management to get maximum exposure at a show or with an EAC to minimize setup costs, remember that if you don't ask, you won't receive.

(This article is edited according to Internet information)

 NEW WORDS AND PHRASES

management　n. 经营; 管理(方式); 管理者/部门

trickery　n. 欺骗, 哄骗, 要花招

sponsorship　n. 赞助; 发起; 倡议; 保证人

adversarial　adj. 敌手的, 对手的, 对抗的

audiovisual　n. 视听设备, 视听教材　adj. 视听的

booth-space　n. 展位空间

rental deposit　押租

exhibit-disposal fees　展品处置费

marshalling yard　铁路货运编组站

waive accessorial charge　免除附加费用

fuel surcharge　燃油附加费

 NOTES

1. EAC（Exhibitor-Appointed Contractor），参展者指定承包商。指除了指定的正式展览承包商之外，任何可以提供服务给参展者的公司的总称。

2. GSC（General Service Contractor），正式的展览承包商或展览服务承包商。具体工作中指的是可以提供会展活动管理和为参展者广泛服务的组织。其对展场的服务包括参展者管理、现场活动协调、摊位装设和拆除、摊位规划、陈列和设计、交通运输服务、物资处理、顾客服务等。其对参展者的服务包括展览整体设计及现场建置、摊位架设和拆除、家具和附属装饰提供、标志或招牌装设提供、物资处理、参展者服务等。

 EXERCISES

Ⅰ. Answer the following questions.

1. What do you need to prepare before negotiating?

2. How to conduct a high-quality negotiation?

3. Which vendors will you deal with in the exhibition service?

4. How to reduce your electrical cost based on your minimum requirements?

5. What services can EAC provide for you?

II. Please translate the following English sentences into Chinese.

1. One of the benefits of doing business in a slow economy is that show management and industry suppliers really want and need your business.

2. The following list represents a variety of the people and suppliers with whom you can negotiate throughout the trade show management process.

3. Many exhibitors limit their interactions with show management to logistical matters such as purchasing booth space.

4. Many exhibit houses charge storage fees based on the amount of cubic feet from the floor of your storage space to the ceiling, regardless of whether your stored items actually occupy all of that available space.

5. I was able to eliminate the exhibit-disposal fees quoted by one exhibit house by getting it to recycle the aluminum from my client's booth.

6. The recycling income offset the landfill fees for the rest of the structure.

7. With talent companies, I've been able to negotiate away the 10-percent agency fee they often charge in addition to the stated daily rate for talent.

8. If you make reasonable requests and build strong partnerships with your suppliers, they will work with you to meet your objectives and your budget.

III. There are 10 sentences in this section. Beneath each sentence there are four words or phrases marked A, B, C and D. Choose one word or phrase that best completes the sentence.

1. Which of the following is the best translation of the phrase "negotiation is not trickery"? (   )

A. 谈判不棘手   B. 谈判不容易   C. 谈判不是诡计   D. 谈判不吹嘘

2. The phrase "get greedy" most probably means (   ).

A. 贪得无厌   B. 自私自利   C. 唯利是图   D. 野心勃勃

3. Figure out your acceptable bottom line ahead of time "bottom line" probably means (   ).

A. base line   B. highlight   C. profit   D. acceptability

4. Use this information as a guideline to help you prepare for your negotiations, the phrase "guideline" is also called (   ).

A. outline   B. proposal   C. route   D. method

5. The phrase "fallback position" is also called (   ).

A. stand point   B. rear position

C. second line position   D. final concession

6. The phrase "split the charges" probably means (   ).

A. sharing fees   B. dividing fees

C. charges sharing   D. denounce a charge

7. Which of the following fees is usually difficult to negotiate? (   )

A. booth-space pricing   B. warehousing fees

C. fees of preparing exhibit for a show   D. turnkey fees

8. Which of the following is not covered by the exhibition venue fee? (   )

A. rental of exhibition venue   B. rental cost of venue facilities

C. exhibition design and layout cost   D. printing cost

9. Which of the following do not belong to "other vendors" in trade show services? (   )

A. audiovisual contractors   B. talent companies

C. customers   D. furniture-rental companies.

10. Which of the following is not an additional cost of negotiation? (   )

A. sponsorship   B. advertising in the show daily

C. speaking opportunities   D. trade order

Ⅳ. Writing

Please write a composition about "How to Win the Business Negotiation of Overseas Exhibition" with the following particulars (No less than 200 words) :

Adopt correct negotiation principles (principle of equality and mutual benefit, principle of seeking common ground while reserving differences, principle of compromise and complementarity).

Adopt appropriate negotiation methods (soft negotiation method, hard negotiation method, principle negotiation method).

Improve the overall quality of exhibitors (good language application ability, keen observation, respect for customer cultural traditions, and superb creativity).

Reading 2

This Old Exhibit: How to Refresh a Tired Booth

Follow these budget-friendly fixes when your exhibit needs a little TLC.

Trade show marketing is a rough-and-tumble business for all involved: exhibit managers, staffers, suppliers, and, of course, the stands themselves. After just a few shows, your shiny new booth can start looking a little rough around the edges due to the normal wear and tear of shipping, installation and dismantle, as well as general use on the show floor. And on top of the cosmetic dings and scratches, your stand can appear a bit tired when attendees encounter the same floor plan and in-booth experiences at multiple shows, creating the impression that your company has nothing new to offer.

In a perfect world, these problems would easily be remedied by a new build. Unfortunately, few exhibit managers are in a position to procure a new exhibit at the drop of a hat. So, since many of us are left to make do with what we have, here are my low-cost tips for extending the life of your exhibit-at least until you have the budget and executive-level buy-in to purchase a new one.

Employ a Little Elbow Grease

White is a popular color in exhibit design, but it is an unforgiving hue when it comes to showing signs of wear. A deep and thorough cleaning can work wonders on a dingy exhibit, and I'm convinced that Mr. Clean Magic Erasers are one of the most miraculous inventions of all time. These little rectangles can quickly and safely tackle scuffs, stains, and marks on countless hard surfaces, turning everything from countertops to storage-room doors from drab to fab. Some exhibit managers also use them on white silicone-edge graphics (SEGs), although I'd advise checking with your supplier before taking one to your own fabrics. The last thing you want to do is turn a small mark in the corner of an SEG into a blotchy mess.

While your sleeves are rolled up, don't overlook easy−to−forget areas, such as the bases of display units and other parts of the exhibit that are close to the ground, which are easily soiled by attendee foot traffic and handling during I&D. Finally, be mindful of transparent elements, e. g. , the windows looking into your enclosed conference room, that would benefit from more than a quick spray of cleaner before show opening.

Swap Out Damaged Surfaces

If your countertop and other flat−surface woes are more than skin deep-think chipped laminates and serious scratches that no amount of scrubbing or stain markers will fix-then the time is right to replace them. This can be a relatively inexpensive remedy, as you're just purchasing the top of a unit, be it a welcome desk or product−display table.

If your exhibit uses shelves of any sort, give these a careful once−over as well. In my experience, shelving tops the list of exhibit components most susceptible to damage. However, like counters and tabletops, it can be easily replaced. But before you spring for an apples−to−apples substitute, consider this a chance to introduce a new color, texture, or material that can make an outsized impact for a small investment. For example, just because the majority of your exhibitory boasts a wood−grain finish doesn't mean you can't add a complementary aesthetic, such as faux concrete or tempered glass.

Recoat Your Extrusions

Many of the aluminum frames used in portable/modular exhibit systems are either powder−coated or treated with an anodized finish. Both offer cosmetic and protective benefits, but deep scratches and gouges are sure to occur over time. Instead of replacing your marred frames with new extrusions, ask your exhibit house or a local metal shop if it can apply a fresh powder coat that will make them look good as new. It may also be possible to have the extrusions finished in a different shade, giving you a budget−friendly opportunity to introduce a refreshing pop of color.

Turn Toward the Light

Effective lighting can make or break any booth, and it can definitely spruce up your timeworn stand by making it look more modern and welcoming. LED arm lights are an inexpensive add−on, and some can be operated by battery pack so you don't need to pay for an additional outlet or worry about complicated wire management. LED puck lights and lighting tape/tubing are also fantastic at adding illumination to hard−to−reach areas, such as shelved displays and other small recesses. Another recent development is the use of floor−mounted multicolored LED units that uplight exhibit walls and other vertical surfaces to create expansive light "washes" capable of changing the ambiance of your entire stand with the push of a button.

Flex Your Creative Muscles

Let's say your logo's been redesigned, the stock photos on your back wall are now bordering on cliché, the colors of your hard−infill panels are dated, and your exhibit−marketing wallet is nearly empty. Rather than despair, put on your thinking cap and look for an outside−the−box solution. One of my clients was once desperate for a refresh and could afford little more than a gallon of paint. So that's exactly what we purchased: a bucket or two of chalkboard paint, to be exact. The result was an engaging, Instagram-worthy backdrop attendees decorated themselves with multiple colors of chalk. I've also seen exhibitors use hook-and-loop fastener to drape beautiful fabrics from their damaged Sintra panels. My point is that from paint to PVC board, there are means of freshening up a stale stand that don't have to break the bank.

Tweak Your Floor Plan

If your exhibit isn't so much worse for wear as you feel stuck in a spatial rut, consider taking a step back and assessing how your stand's components can be repositioned to create a new look. This could be as simple as shifting a few modular elements to totally rethinking the layout using the exhibitory you already have. For example, what if you moved your reception desk from its front-and-center location to an angled position at the corner? Or perhaps the freestanding L-shaped pop-up wall you normally place near the aisle could be moved to the middle of the booth to create a dedicated area to showcase a particular product or service. Depending on the complexity of your stand design, it may be best to work with your exhibit house to identify all the available options, but you'd be surprised how rearranging a few elements can result in a like-new look and feel.

On a similar note, if the "bones" of your exhibit are solid but you worry that you're boring attendees with the same-old, same-old displays, look for ways to increase engagement by incorporating an interactive experience, such as a hands-on product demo or in-booth game that reinforces your company's marketing messages. In some cases, making space for an experiential activation may necessitate scrapping one or two of the humdrum, static elements that booth visitors have seen before, but their presence will hardly be missed by attendees busy learning about your goods or services in a dynamic, memorable way.

After reading these tips, those of you who are thinking of purchasing a new exhibit may be wondering what steps you should take to ensure the longevity of your upcoming investment. First, I highly recommend opting for modular exhibit components whenever possible, as these will allow you to make incremental repairs and updates to your exhibit as needed. Second, many reputable suppliers offer warranties on items such as hanging signs and pop-up banners, so give your dollars to a company that stands behind what it sells.

At the end of the day, as face-to-face marketers, our booths are representative of our respective brands. It is therefore essential for us to make them the best they can be, regardless of our abilities to spring for brand-new designs and components whenever it suits us. But armed with this advice, you'll surely be able to prolong the life of your show-floor workhorses-provided you're also armed with an ample supply of creativity and Magic Erasers.

(This article is edited according to Internet information)

 NEW WORDS AND PHRASES

rough-and-tumble   adj. 杂乱无章的, 混乱的

dismantle   vt. 拆开, 拆除; 废除, 取消

dingy   adj. 暗淡的, 乏味的; 肮脏的

countertop   n. (厨房的) 工作台面

swap   n. 交换; 交换物

silicone-edge graphics   硅胶边缘图形

laminate   n. 层压材料; 叠层, 层压

exhibitory   adj. 展览的

wood-grain   n. 木纹

recoat   vt. 再涂在上面, 重新涂

extrusion   n. 挤出; 推出; 喷出; 赶出

modular   adj. 组合式的; 模块化的

mar   vt. 毁坏, 损坏; 弄糟; 糟蹋; 玷污

spruce up   打扮整齐

puck lights   冰球灯

lighting tape   照明胶带

cliché   n. 陈词滥调

Sintra panel   辛特拉面板

spatial rut   空间凹槽

reposition   v. 使复位; 改变……的位置

static element   静态元素

 **Reading 3**

Downsizing

Downsizing is de rigueur these days, as exhibitors reassess their footprints and reconsider exhibit components to make room for more social distancing. So, exhibitor recruited a panel of experts to help you reduce real estate and exhibitory without ruining effectiveness.

As in-person trade shows dust off the cobwebs and sputter to life, chances are that many exhibitors will downsize their exhibitory to one degree or another. While some will reduce their footprints due to decreased marketing budgets and fewer attendees and staff traveling to shows, others will maintain their square footage but scale back on the amount of exhibitory employed within it. The latter provides more space for social distancing and COVID-related adaptations, plus it can cut costs, as less exhibitory typically translates to lower transportation, drayage, and labor fees. What's more, if a show assigns priority points to exhibitors with larger footprints, maintaining a sizable slab of concrete may pay off in the long run.

The point is, when most exhibitors hit the road, they'll likely need to leave something or someone at home that would have previously made the trip. But having never been down this precarious path before, most marketers aren't quite sure what to bring nor what to eliminate. After all, they don't want to overpack, but they can't leave crucial components at home either.

That's why exhibitor tapped a host of exhibit-design and-marketing experts for their downsizing advice. Along with tips to help you scale back components and afford more social-distancing space, they've provided several queries to aid the process along with specific "ditch that" directives. Their guidance should help you tighten your belt without decreasing effectiveness.

Investigative Questioning

The following queries can help determine whether specific exhibit elements make the trip or stayat home.

Unfortunately, you can't just toss components into a booth space and call it a marketing medium. Rather, you must choose exhibitory based on marketing objectives and attendee needs. The thing is, those criteria may have changed dramatically since COVID became a thing. As you launch your downsizing endeavors, experts suggest asking yourself the following.

- What's our No.1 goal?

"Identifying your one big thing helps you understand what must be included and what might be unnecessary," says Jay Menashe, CTSM, vice president of sales and marketing at EDE Corp. For example, if your primary aim is to launch a new gizmo, you'd prioritize demo stations, and if push came to shove, you'd eliminate conference spaces, awareness-related elements, etc.

- What do we want visitors to remember?

Another way to isolate your core reason for exhibiting-and therefore ascertain the essential exhibitory that supports it-is to identity the sole thing you want visitors to remember about their experiences in your stand, says Katina Rigall Zipay, creative director at Classic Exhibits Inc. "Often, the memory you want to instill is the name of a new product, a key message, or a positive impression of your staff or the brand," she says. "But by pinpointing this aim, you can also make sound decisions about your exhibitory."

- How many staff and attendees do we expect?

In June 2020, the International Association of Exhibitions and Events (IAEE) recommended an optimum floor density (based on a 6-foot radius around individuals) of 28 square feet per person. Based on this suggestion, those with a 20-by-20-foot booth, for example, would limit occupancy to 14 people. While recommendations are changing rapidly, each show also may have its own guidelines regarding this metric. The point is, it's important to determine how many attendees will occupy your space at once and how many staff you intend to bring to the show. "This data can help you determine the amount and type of exhibitory you need," Menashe says. "For example, if only three people will occupy your exhibit due to density recommendations, do you really need a reception counter? And if you're bringing only four staffers, do they need a sizable storage or break area?"

- What do attendees need?

In all likelihood, attendees' pain points and expectations have changed in the last year or two. "That means you can probably forget some of what you knew about your audience," says Melissa Park, founder of Melissa Park Events. "Therefore, it's time to ask some hard questions to reassess each show's unique attendance. Who's going to this show today, not two years ago? What do they need, and how have these needs changed since 2019? How can your offerings and your booth best meet their expectations?" Stephen Ross, vice president and executive creative director at Access TCA Inc., further stresses the importance of an attendee-centric experience. "To create such an encounter, determine how you want visitors to feel and what you want them to think and know about your brand or product," he says. "Also consider how you want them to act after engaging with you. If you understand your audience in these terms, you can choose components that will facilitate face-to-face conversations."

- How will staff engage with attendees?

The whole point of trade shows is to allow your personnel to talk with attendees face to face. "Always keep visitor engagement and interaction top of mind as you prioritize exhibit planning," says Bill Lanisek, event strategist from Novak Birch Inc. Obviously, pandemic protocols mean this engagement will look a little different than it did in 2019. "But to effectively downsize exhibitory," he says, "you need to understand and prioritize elements that best support face-to-face engagement results." Along these same lines, Todd Dailey, vice present of creative at Visual Communications, reminds us that attendee "returns" are also critical. "The new normal for exhibiting is all about focusing your message to attendees' pain points," he says. "Gone are the days when an errant salesperson can shoehorn in one extra workstation for a specific product or service that has a small effect on your business. Today it's about what is important to your audience and what gives them the best return on experience."

Going Big in a small space

How do you ensure that your petite environment still looks spacious and inviting as opposed to cluttered and cramped? Our experts weigh in.

- Declutter. Clutter kills any booth and can deter attendees from entering your space, says Bill Lanisek, event strategist at Novak Birch Inc. So whenever possible, maintain a tight focus on your main messages, goals, and products, and eliminate everything else.

- Avoid dark colors and patterns. Designers often use patterns and dark colors to elicit emotions and intimacy, says Jay Menashe, CTSM, vice president of sales and marketing at EDE Corp. But if you want to maintain a sense of openness within a small space, bypass the patterns and aim for lighter colors.

- Incorporate some white space. Dense graphics perceptually diminish the size of your stand. "It's important to maintain some white space and not overfill your graphics with messages or text," says Katina Rigall Zipay, creative director at Classic Exhibits Inc.

- Maintain you sightlines. "Attendees want to be able to see through and into your space, so designing clear, uninterrupted sightlines to key points of interest is paramount," says Sean Combs, CEO of Steelhead Productions. See-through walls, slatted dividers, and more are a few ways make a smaller space appear larger.

Unnecessary Accoutrements

Take a close look at these frequently used exhibit elements, which are probably due for a decrease. While the previous queries will jumpstart the downsizing process, experts offered further advice for culling components. Here are a handful of elements that are ripe for reduction.

- Enclosed meeting rooms. Face-to-face meetings are almost always essential, so you'll need to set aside pandemic-friendly environs for these endeavors. However, Sean Combs, CEO of Steelhead Productions, says it may be time to skip enclosed conference rooms for the near future. "Many attendees won't be eager to enter such an environment-or any close quarters for that matter.

A simple solution is to eliminate traditional closed-door conference rooms from your floorplan."

● Paper and digitizable assets. Eschew paper collateral (e. g. , brochures, product one-sheets, etc.) along with any video assets that don't clearly support your main directives. "If people can scan a Quick Response (QR) code and gain access to the content immediately and/or retrieve it from the comfort of their homes or offices-then you can eliminate the need for literature stands, collateral storage, and even a few monitors and perhaps the walls or cabinetry that support them, " says Kevin Carty, executive vice president of Classic Exhibits.

● Hospitality and gift items. Dailey expects that some attendees may not feel 100-percent comfortable with food service, so prepackaged items may be the best way to go. Lanisek concurs. "You may want to reconsider-or at least redefine-hospitality and giveaway offerings, " he says. "Assess the need for self-serve coffee stations, candy bowls, and any kind of food or tchotchke distribution." Granted, you could switch to single-serve, individually wrapped items, but first ascertain whether a particular offering truly supports your objectives.

● Lounge furniture. While you may want to sit down for socially distanced discussions, presentations, and demos, Lanisek recommends that you eliminate most lounge furniture that might entice visitors to loiter in your space. You'll likely need to curtail the number of people in the exhibit at any given time, and you don't want unqualified and unengaged attendees simply lolling about.

● Nonessential or underqualified staff. No, people certainly aren't exhibitory. But they-and their stuff-still take up space. "When selecting booth staff, consider who will best serve the needs of both the company and the attendees, " Menashe says. "Now's the time to bring your starting five (or two or 10) and to leave the benchwarmers at home." Also consider setting up a sort of hybrid staffing team. That is, you could bring a few of your top-notch salespeople but then have several product specialists and executives on standby for impromptu or scheduled booth meetings via Zoom.

● Heavy, cumbersome components. Scrutinize all your burdensome elements, Zipay suggests. "These pieces often cost a pretty penny in terms of shipping, drayage, and installation, and if they don't pack down well, they may take up more room in your crates, " she says. "So carefully look over such components to determine if they're truly critical to your objectives or if you can leave them at home and kill two birds with one stone-by freeing up space and decreasing costs."

Reduction Instructions

Further enhance your downsizing efforts by fully leveraging the components you do bring to the show.

Downsizing a design involves far more than merely purging some pedestals. You also need to make smart choices about the remaining exhibitory and leverage the keepers to their fullest extent. Here are some tips to do just that.

● Employ dual-function elements. You've likely seen reception desks that act like Swiss army knives with built-in demo screens, and storage space, but myriad other elements can multitask. "The same wall or kiosk could have an interactive screen on the front and a product display on the side while its back side defines a meeting space, " Zipay says. Even something as simple as a strategically placed bench can delineate a space and direct traffic.

● Replace on-site swag with an online store. "A great alternative to in-booth giveaways is a virtual swag store filled with branded items or gift-card codes attendees redeem online," Park says. "Qualified visitors can view the store via a tablet (which you can sanitize after each use), input their contact info, and pick what they want. After the event, you simply email or mail them their selections."

● Reduce, don't remove. Rather than completely doing away with a piece or type of exhibitory, consider how you might edit it. "Does your storage closet need to be as big as in years past, particularly since you're likely bringing fewer employees and handing out less swag?" Menashe says.

● Consider traffic flow. To make the most of your selected exhibitory, ask show management about any COVID-induced traffic flow changes or show-floor tweaks that could affect how you orient your booth, Carty says. For example, you don't want to position your massive LED display in the opposite direction from which attendees will be approaching your space.

● Keep effectiveness top of mind. Whether your downsizing endeavors are driven by the pandemic or something else entirely, Park cautions that effectiveness trumps everything. "If you can downsize your space and deliver a great experience that meets your goals and objectives, then you absolutely should," she says. "But if you can't deliver an experience that represents your brand or showcases your products in the best light, then that event could be more detrimental than beneficial."

(This article is edited according to Internet information)

NEW WORDS AND PHRASES

COVID-related   adj. 与新冠病毒相关的

overpack   v. 过度包装

gizmo   n. 小发明

prioritize   v. 按重要性排列, 划分优先顺序; 优先处理

International Association of Exhibitions and Events   国际展览与活动协会

attendee-centric   adj. 以与会者为中心

spacious   adj. 宽敞的; 广阔的; (知识) 广博的

dense graphics perceptually   密集的图形感知

tchotchke   n. 廉价小饰物; 小玩意

ENGLISH FOR WORKPLACE COMMUNICATION

Sample Dialogue: Enterprises participating in exhibitions

Situation: The following conversation is between Tom, a manager from ABC company(B), an employee named Jack (A), they are talking about enterprises participating in exhibitions.

A: Hello, manager, I am here for a consultation regarding enterprises participating in exhibitions.

B: Yes, please.

A: What are the purposes of enterprises participating in the exhibition?

B: Probably to understand new markets, look for import and export opportunities, exchange experience and understand development trends. What's more to understand the status of the company's industry, seek cooperation opportunities.

A: How do enterprises choose exhibition companies with good reputation?

B: The staff of the enterprise can put forward some problems related to the exhibition. If the exhibition company gives targeted suggestions to the enterprise, it can help the enterprise to analyze the interests and consider for the enterprise. This shows that the staff of the exhibition company or organization are very professional.

A: Many enterprises have concerns "Can the cost be recovered after the exhibition"?

B: The role of the exhibition is to build a platform for buyers and sellers to meet on this platform. As for whether the business can be completed, there are many decisive factors.

A: Why are there different quotations for the same exhibition?

B: The quotation varies with the location of the booth and the decoration scale of the booth. Enterprises should know what the decoration of the booth includes, not just the quotation, but the cost performance ratio of the quotation.

A: Is the continuity of enterprise participation important?

B: For the first exhibition, enterprises should not pay too much attention to the effect. The effect depends on many factors, such as whether the product has a market, how about the product quality, and the composition of target customers. Enterprises should have a good attitude when going abroad to participate in exhibitions, and the participation must be continuous.

A: All right, thank you very much!

B: It's my pleasure.

BASIC KNOWLEDGE ABOUT TRADE FAIRS

本章导读

　　本章介绍了国际展会的一些基本知识，包括展位、展馆及展品运输。通过本章的学习，了解展位的分类，不同展位类型在展馆中的位置，各自的优势及劣势，学会如何根据参展公司的自身特点，有目的地选择适合自己的展位。在一场展览会上，展位的类型和位置会影响参展企业的观众流量和展出效果。比如，整岛展位能充分展示一个公司的实力，双开（角）展位位于两条通道的交口，是标准展位中的首选。学习展馆规则，特别是对展位搭建、现场管理等方面的具体要求。展品运输是保障展会顺利举办的重要环节，了解展品运输的特点，学会如何选择信誉良好的物流公司来承运展品等。

知识精讲

展会知识

从服贸会看电子商务行业新亮点

　　新华社北京 9 月 3 日电（记者宋玉萌、吉宁）在 2022 年中国国际服务贸易交易会（以下简称"服贸会"）上，商务部发布《中国电子商务报告（2021）》，直观地展示了电子商务产业的蓬勃动力。作为数字经济最具创新和活力的领域之一，电子商务在服贸会期间呈现出哪些新亮点？

绿色低碳凸显社会责任

　　数字化技术正推进消费全周期、全链条、全体系深度融入绿色理念。许多电子商务企业围绕"双碳"目标提升社会责任，在物流、运输、供应链等多方面为绿色发展提供持续动力。京东零售 CEO 辛利军表示，将供应链作为核心能力的京东，2022 年提出了"有责任的供应链"的概念，即从原来的商业价值、产业价值，进一步延伸到社会价值。

　　据介绍，2021 年京东物流在海口市进行循环包装试点项目，通过投放循环快递袋探索逐步减少使用一次性 PE 塑料包装袋。此外，截至 2021 年年底，京东物流已在全国 50 多个城市投放了 2 万辆新能源物流车，能实现每年约 40 万吨的二氧化碳减排。

　　生活服务类企业也在服务领域持续推动绿色低碳工作。"本地生活行业着力加大在促消费、促城市发展、促美好生活三个维度上的服务力度。"饿了么总裁方永新介绍，近年来饿了么已在"三保""三促"上持续开展多项创新，比如消费者下单时可以勾选是否需要餐具、小份菜等。

直播+助农成为经济新引擎

根据《中国电子商务报告（2021）》的数据统计，2021年全国农村网络零售额2.05万亿元，同比增长11.3%。全国农产品网络零售额4 221亿元，同比增长2.8%。农村正在成为电子商务发展新的增长极。

一批企业借助服贸会的展台也展示了电子商务服务农业的最新成果，农产品直播电子商务平台东方甄选今年首次亮相服贸会，将直播大屏搬到现场，让观众近距离感受"知识直播"和农产品直播带货的独特魅力。

新东方教育科技集团副总裁尹强表示，农产品是非标品，品控难、运输难，在电子商务行业中具有极强挑战性，但拥有十分广阔的市场空间。"一方面，直播电子商务可以帮生产者解决不善于做品牌、优质产品打不开销路的难题，帮农产品工厂搭建销售渠道、打造品牌，更有机会激活当地经济，盘活产业链。另一方面，从消费者角度，人们对美好生活的向往和追求让大家对农产品品质的要求变得更高。"他说。

直播电子商务不但走出助农之路，也在国际上打出知名度。服贸会的文旅板块组织了相关活动，为参展各国驻华使馆提供了线上宣传推介平台，首次设置"一带一路"沿线各国特色文化产品"国际甄选"电子商务直播间，打造"国家场"环节。展期内每天都有"一带一路"沿线国家外交官、企业负责人"带货直播"，介绍各自国家的特色产品。

跨境电子商务带动产业发展新方向

商务部相关负责人在服贸会跨境电子商务论坛上介绍说，跨境电子商务成为助力稳住外贸基本盘、推动外贸转型升级和高质量发展的重要抓手和重要动能。

目前一批垂直类及平台型电子商务企业纷纷深度布局跨境电子商务业务。来自北京的小笨鸟跨境电子商务平台，通过整合跨境供应链服务企业及服务资源，为国内出口企业提供端到端的整体服务，并在美国、欧洲、中东建立了海外分支机构和海外仓，形成了线上线下相融合的数字化跨境贸易供应链综合服务体系。

"中国品牌已经开始重新定义全球品牌营销的玩法，将借助新的机遇进一步发展。"小笨鸟跨境电子商务平台高级副总裁张革认为，目前政策机遇、数字经济机遇和产业链机遇三方面因素叠加，为品牌出海带来很好的机遇期。

中国国际电子商务中心电子商务首席专家李鸣涛指出，中国企业通过跨境电子商务助力做好自己的核心业务，有助于企业竞争及全球服务能力提升。

（本文根据互联网新闻报道整理）

思考并回答以下问题：

1. 服贸会的参展商通常都是什么类型的公司？服贸会与广交会有哪些不一样？

2. 服务贸易涵盖哪些内容？

3. 什么是跨境电子商务，你了解到的跨境电子商务平台都有哪些？

4. 为什么我国近年来大力发展跨境电子商务？

5. 直播带货也是展览展示的一种方式，近年来非常火爆，很多人转向抖音等直播平台购物。新东方直播带货诞生了新网红，你喜欢董宇辉的直播吗？为什么？

Reading 1

Standard USA Booth Regulations and Types

Did you know there are actually four different types of booths you can create for trade shows? Booths have different characteristics that involve how they are positioned on the trade show floor and accessible, as well as the allotted space. Below we cover the exact differences between the four types of trade show booths, a few quick tips to maximize usage of each booth type, and some great examples of each booth in action.

1. Island Booth

Island booths are open to show-floor aisles on all four sides, meaning the interior is often highly visible regardless of which direction you approach it from. Booth dimensions range, but they tend to be 20 feet by 20 feet or larger. The size of your island is typically only limited by budget or show restrictions, with 5,000 to 10,000-square-foot spaces being more common than some might think.

Use an island booth when you want to maximize visual space. You have the opportunity to include marketing messaging on all sides, and there are typically fewer restrictions around hanging signs compared to the other types of trade show booths. You also gain the flexibility to determine how attendees will come into, move about, and exit your space. Not to mention there's more breathing room between your booth and those of the other exhibitors.

Quick tips to make the most of an island booth:

● Each side of the booth is highly visible, so be sure to take advantage of that. You never know where an attendee will be approaching from.

● If the island is large, incorporate enough free space to avoid overcrowding, which will improve the attendee journey.

Island Booth Example: Scorpion (at Service World) (Figure 9. 1)

Figure 9. 1

This 30-by-30 island booth was constructed using shipping containers, which comprise the majority of its architecture. Scorpion's Branded Environment takes advantage of the flexibility of an island through easy access on three sides, with stairs leading up to the second level tucked away in the back.

Island Booth Example: Misfit (at CES) (Figure 9. 2)

Figure 9. 2

This 20-by-40 island booth makes use of all four aisles to display products and offer attendees different ways into the exhibit. Since CES is such a busy and well-trafficked show, it was important for Misfit to be highly visible and easily accessible.

2. Peninsula Booth

Peninsula booths have aisles on three sides, and they share a wall with another exhibit. Think of the peninsula as the middle child—it tends to be smaller and less expensive than an island but more expensive than perimeter and linear booths. Dimensions are typically 20 feet by 20 feet or moderately larger.

If you want some flexibility without the higher price tag of an island, a peninsula booth is a great choice. You may not have enough messaging or materials to warrant four sides, so you can invest more in the rest of the space.

Quick tips to make the most of a peninsula booth:

• Know the rules and regulations regarding the exterior wall you share with the other booth. Typically, the wall must be plain and devoid of branding. If you have to make changes to comply, it can be challenging, costly, or visually unappealing—some shows force you to use unattractive piping and drapes.

• Like the wall, find out in advance what can and cannot go on any hanging signage facing the exterior wall.

• Ensure the attendee journey can be accomplished within the three-sided configuration—

sometimes the missing aisle/entrance can throw off an otherwise well-designed activity.

Peninsula Booth Example: High Sierra (at Outdoor Retailer's Summer Market) (Figure 9. 3)

Figure 9. 3

This 30-by-40 peninsula booth uses walls on two aisles to funnel attendees to the front, ensuring they start at the reception desk. High Sierra products line the walls, and there are a couple of meeting areas along the back wall, enabling staff to sit down and have private discussions with visitors and clients.

Peninsula Booth Example: Khombu (at Outdoor Retailer's Winter Market) (Figure 9. 4)

Figure 9. 4

Khombu used a 20-by-20 peninsula with an open, asymmetrical layout to grab attendees from two aisles. The back wall that faces the adjacent exhibit extends upward 12 feet to ensure the Khombu brand can be seen from a distance—important since the Branded Environment features no hanging signage.

3. Perimeter Booth

Perimeter booths are similar to peninsulas—one side will not be accessible—except these booths are on the outer perimeter of the trade show floor. Perimeter booths can be L-shaped if in a corner, meaning their dimensions often skew away from the traditional square—they start at 10 feet by 10 feet, and often come in 10-by-20 arrangements.

You may consider a perimeter booth if you want to maximize your vertical visibility, as these booths tend to have relaxed height restrictions. You also don't have to worry about the back wall since the space beyond it isn't walkable.

Quick tips to make the most of a perimeter booth:

• Be aware of whether your space is sharing a wall with an exhibit; if so, you'll need to know the rules and regulations on wall-sharing (like with a peninsula).

• Make use of the perimeter's unique dimensions—they may even influence your design concept.

• Confirm that the design you come up with works comfortably in the space.

Perimeter Booth Example: Scorpion (at MTMP) (Figure 9.5)

Figure 9.5

This 10-by-40 booth is both a perimeter and a linear type. Through efficient use of space, it features two seating areas, a presentation area, and a reception area. Dramatic, refined architecture and lighting details set this Branded Environment apart from everything else around it.

Perimeter Booth Example: Super. natural (at Outdoor Retailer's Winter Market) (Figure 9. 6)

Figure 9. 6

Super. natural employed a simple 20−by−30 perimeter booth nestled in the corner of the exhibit hall. It takes advantage of the placement by creating an open and accessible space. The square hanging sign is unique in its size, reflecting the proportions of the brand's logo.

4. Linear Booth

Linear booths are the most common types of trade show booths. They are mostly accessed by one show aisle—or two if located in a corner. Linear booths share at least two walls with other exhibits. The most common dimensions are 10 feet by 10 feet, but they can be slightly larger.

The linear booth is the least expensive option. So, if it's important that you have a presence at a show but you're also budget−conscious, linear is the way to go. Just keep in mind these booths have the most limitations—limited space, limited accessibility, limited visibility, and so on. Plus, an enclosed space tends to feel smaller.

Quick tips to make the most of a linear booth:

- Be aware of height restrictions because you typically can't go as high as other booth types.
- Use considerate design to make the best use of the space.
- Staff sparingly—with so little space, you don't want to overpopulate the booth.

Linear Booth Example: Mapbox (at AWS) (Figure 9. 7)

Figure 9. 7

This is a 10-by-20 linear booth flanked on two sides by neighboring exhibits. It features two **presentation** areas on the left and right, and a large video tile screen in the center. The video **screen** was used to display the brand's map technology in dramatic fashion.

Linear Booth Example: Threat Connect (at RSA) (Figure 9. 8)

Figure 9. 8

Threat Connect's 10-by-10 linear booth used bright graphics to stand out amongst the competition. The space has a simple back wall with a small closet and screens, but the elements work well together and give the brand great presence on a crowded show floor. On the right side is an example of a divider wall of piping and drapes.

From the above we know that there are four types of booth configurations: Standard/Linear,

Perimeter Wall, Peninsula, and Island. The following booth display rules are typical for U.S. Trade Shows and Convention Halls. However, regulations vary by convention center and even within show halls. Contact show management for specific regulations.

➤ Standard/Linear Booth (10′ depth, 9 square meters)

Any booth that shares a common back wall and abuts other exhibits on one or two sides. Maximum height is 8′. This 8′ height may be maintained on the sidewall of your booth up to a distance of 5′ from the front aisle. The remaining length of the sidewall may be no higher than 4′. A corner booth is a linear booth exposed to aisle on two sides. All other guidelines for linear booths apply.

Note: Hanging signs are not permitted over standard/linear booths.

➤ Perimeter Wall Booth (10′ depth)

A standard/linear booth found on the perimeter walls of the exhibit floor. The maximum height is 12′. This 12′ height may be maintained on the sidewalls of your booth up to a distance of S' from the front aisle. The remaining length of the sidewall may be no higher than 4′.

Note: Hanging signs are not permitted over perimeter wall booths.

➤ Peninsula Booth

Any exhibit 20′×20′ or larger with a depth from the common back wall to the aisle of at least 20′ and with aisles on three sides. There are two types of Peninsula Booths: (a) one that backs up to Linear Booths, and (b) one that backs up to another Peninsula Booth and is referred to as a "Split Island Booth." For all peninsula booths, the exterior of the back wall must be plainly finished and may not contain booth identification, logos or advertisements.

If backed by a row of standard/linear booths, the back wall may be no higher than 4′ for a distance of 5′ from either side aisle and 20′ high in the center of the back wall. These height restrictions must be maintained for a distance of 10′ from the back wall. Where two (2) peninsula booths share a common back wall ("split Island"), the maximum height may be 20′ in all areas of the booth, including the back wall (same as Island Booth rules, below).

Note: Hanging signs are permitted over peninsula booths that are 20′×20′ or larger.

➤ Island Booth

Any exhibit 20′×20′ or larger and is surrounded by aisles on four sides. Regulations vary by exhibit hall but the following are typical examples: Maximum height of 30′ in all areas of your booth allowed in North Hall and Central Halls 3−5. Maximum height of 20′ in all areas of your booth is allowed in Central Halls 1−2. Maximum height of 22′ in all areas of your booth allowed in South Halls. No limitations on the number of solid walls for your Island booth. Be sure to check the hall regulations.

Note: Hanging signs are permitted above island booths.

OTHER CONSIDERATIONS

➤ Canopies and Ceilings

Canopies, including ceilings, umbrellas and canopy frames, can be either decorative or functional (such as to shade computer monitors from ambient light or for hanging products). Canopies for Linear or Perimeter Booths should comply with line-of-sight requirements.

The bottom of the canopy should not be lower than 7 ft. (2.13 m) from the floor within 5 ft.

(1.52 m) of any aisle. Canopy supports should be no wider than three inches 7.62 cm. This applies to any booth configuration that has a sight line restriction, such as a Linear Booth. Fire and safety regulations in many facilities strictly govern the use of canopies, ceilings, and other similar coverings. Check with the appropriate local agencies prior to determining specific exhibition rules.

Covered ceiling structures or enclosed rooms, including tents or canopies, shall have one smoke detector placed on the ceiling for every 900 square feet.

➤ Hanging Signs and Graphics

Hanging signs and graphics are permitted upon approval in all standard Peninsula, Island and Split Island Booths, with a maximum height of sixteen feet (16 ft.) (4.87 m) to the top of the sign as measured from the floor.

Whether suspended from above or supported from below, they should comply with all ordinary use-of-space requirements (for example, the highest point of any sign may not exceed the maximum allowable height for the booth type). Double-sided hanging signs and graphics shall be set back ten feet (10 ft.) (3.05 m) from adjacent booths and be directly over contracted space only.

➤ Theatrical Truss and Lighting

Ceiling-supported theatrical truss and lighting are permitted in all standard Peninsula, Island and Spilt Island Booths to a maximum height of twenty feet (20 ft.) (6.1 m) where ceiling permits. Ground-supported truss may not exceed the maximum allowable height for the booth type. Logos or graphics are not permitted over the sixteen-foot (16 ft.) (4.87 m) height restriction and must have four feet (4 ft.) (1.22 m) of separation from the top of the sign to the top of the truss.

Exhibitors should adhere to the following suggested minimum guidelines when determining booth lighting:

No lighting, fixtures, lighting trusses or overhead lighting is allowed outside the boundaries of the exhibit space.

Exhibitors intending to use hanging light systems must submit drawings for approval by the published deadline date.

Lighting must be directed to the inner confines of the booth space. Lighting must comply with facility rules.

Lighting which is potentially harmful, such as lasers or ultraviolet lighting, should comply with facility rules and be approved in writing by exhibition management.

Lighting that spins, rotates, pulsates, and other specialized lighting effects should be in good taste and not interfere with neighboring Exhibitors or otherwise detract from the general atmosphere of the event.

Reduced lighting for theater areas should be approved by the exhibition organizer, the utility provider, and the exhibit facility.

(This article is edited according to Internet information)

 NEW WORDS AND PHRASES

allotted   adj. 分配的

dimension   n. 空间, 大小

incorporate   vt. 留出, 体现

tuck away   隐藏

well-trafficked   adj. 客流量大

peninsula   n. 半岛

devoid   adj. 全无的

configuration   n. 布局, 结构

funnel   v. 引导

asymmetrical   adj. 不对称的

adjacent   adj. 邻近的

perimeter   n. 周边, 外缘

skew away   偏离

vertical   adj. 垂直的

overpopulate   vt. 使人口过密

canopy   n. 遮篷

ceiling   n. 吊顶

ambient   adj. 周围的

theatrical truss   剧场桁架

adhere   vi. 遵守, 遵循

fixture   n. 设备, 固定装置

laser   n. 激光

ultraviolet lighting   紫外线灯

spin   v. 旋转

pulsate   vi. 有规律地跳动

 EXERCISES

Ⅰ. Answer the following questions.

1. What types of booth configurations do you get from this article?

2. What are the advantages of an island booth?

3. Please describe the peninsula booths.

4. What do you need to pay attention to if you want make the most of a perimeter booth?

5. What are the disadvantages of a linear booth compared to other three types of booths?

6. What are the regulations on canopies and ceilings?

Ⅱ. Please translate the following English sentences into Chinese.

1. There are typically fewer restrictions around hanging signs compared to the other types of trade show booths.

2. You may not have enough messaging or materials to warrant four sides, so you can invest more in the rest of the space.

3. Sometimes the missing aisle/entrance can throw off an otherwise well-designed activity.

4. You also don't have to worry about the back wall since the space beyond it isn't walkable.

5. Be aware of whether your space is sharing a wall with an exhibit.

6. Confirm that the design you come up with works comfortably in the space.

7. Linear booths are the most common types of trade show booths.

8. So, if it's important that you have a presence at a show but you're also budget-conscious, linear is the way to go.

9. Use considerate design to make the best use of the space.

10. However, regulations vary by convention center and even within show halls.

Ⅲ. There are 8 sentences in this section. Beneath each sentence there are four words or phrases marked A, B, C and D. Choose one word or phrase that best completes the sentence.

1. When you want to maximize visual space, which type of booth is suitable? (　　)

A. island booth   B. peninsula booth

C. perimeter booth   D. linear booth

2. If you want to spend the least amount of money, which type of booth can you choose? (　　)

A. island booth   B. peninsula booth

C. perimeter booth   D. linear booth

3. If you want some flexibility without the higher price tag of an island, which type of booth is a good choice? (　　)

A. island booth   B. peninsula booth

C. perimeter booth   D. linear booth

4. Which one is the most common type of trade show booth? (　　)

A. island booth   B. peninsula booth

C. perimeter booth   D. linear booth

5. How many sides do the peninsula booths have aisles on? (　　)

A. one   B. two   C. three   D. four

6. If you want to maximize your vertical visibility, which type of booth may you consider? (　　)

A. island booth   B. peninsula booth

C. perimeter booth   D. linear booth

7. Logos or graphics are not permitted over the (　　) height restriction.

A. 6.1 m   B. 3.05 m   C. 4.87 m   D. 2.13 m

8. The bottom of the canopy should not be lower than (　　) from the floor.

A. 6.1 m   B. 3.05 m   C. 4.87 m   D. 2.13 m

Ⅳ. Writing

Please write a trade show invitation to Mr. Smith, with the following particulars:

1. You are the sales manager of A silk company.

2. The trade show will be held from October 15th to 19th in Guangzhou.

3. Ask Mr. Smith to reply before October 5th.

4. Hope Mr. Smith can accept the invitation and your company will arrange the accommodation.

📖 **Reading 2**

Rules for Success: Trade Show Regulations for a Smooth-Running Exhibition

Trade show exhibitors know the challenges of planning, shipping, and executing a booth exhibit without a hitch. It's no simple feat. Exhibitors and their employers spend significant time, money, and effort planning out each and every detail of the exhibit, with high expectations of a profitable return-on-investment for their efforts.

One thing experienced exhibition managers understand is that there are lots and lots of rules that must be followed before, during, and after every show. There are so many rules in fact, that it can be easy to forget what they are, or which ones apply specifically to them. For rookie exhibitors who are new to the trade show industry, such exhibition regulations can be virtually impossible to remember.

To layout the guidelines clearly, there are multiple sources to be used as resources when planning for an upcoming exhibit. A trade show's regulations can be found in the prospectus, terms and conditions, booth-space rental contract and the exhibitor services manual. Keeping these documents on hand provides exhibitors with everything they need to know to have a great trade show.

To make it a bit simpler for any newbie exhibitors, or for those seasoned ones who just have a hard time keeping all of those regulations in check, here's a list of some of the most important guidelines to remember.

➤ Outboarding and Suitcasing Are NOT Allowed

If you're unfamiliar with these terms, they refer to sales attempts and meetings that occur when and where they aren't supposed to, and by whom they aren't supposed to. Suitcasing occurs when outside businesses or individuals not associated with the trade show try to attract business from show attendees. Outboarding occurs when companies try to hold separate events or meetings outside of the exhibition hall to draw attendees away from the show floor.

They may attempt to use nearby hotel event or meeting rooms to lure prospective clients away from the trade show. If you think that you can draw business at these shows without going through the proper channels and paying the same fees as everyone else, think again. This is greatly frowned upon and trade show managers will do everything possible to deter such actions from occurring.

➤ Exhibitor Trade Shows May Have Individual Rules to Follow

These rules may include but are not limited to: no helium balloons on the exhibition sales floor, no duct tape allowed, or limitations on noise levels at each booth. Exhibitors may even be restricted from using their own electrical cords or power strips. Will you be required to supply carpeting for your booth? And be sure to check on the show's rules regarding in-booth contests or prize drawings. The rules will likely vary from show to show.

Be sure to look at each trade show's rental contracts and terms & conditions to learn which rules you need to be aware of. Neglecting to follow these exhibition rules could result in hefty fines, or get you banned from future shows.

➤ Stay in Your Booth

Known as a "confines of booth" clause, this part of your booth-rental contract restricts you from formally conducting any marketing or promotional activities outside of your designated booth space. Don't plan to roam all over the convention hall trying to hit up attendees with pamphlets, promo SWAG and other materials.

➤ Keep Your Exhibit Dimensions within the Regulatory Standards of the Trade Show

Depending on the venue, there will be height regulations that your exhibit must follow. Whether your booth is labeled an in-line exhibit, perimeter in-line, or island exhibit will determine exactly how high your booth design elements can be. These height restrictions can vary dramatically, so look at your informational resources to determine the exact specifications allowed.

These dimensions also fall under the "line-of-sight" rule of thumb. Your fixtures and promotional features cannot obstruct the view of other exhibits. Trying to break this rule could cost you time and money trying to get your booth corrected before the next days' events.

➤ Federal Regulations Apply at Trade Shows

Specifically, this rule exists in reference to the Americans with Disabilities Act (ADA) of 1990. This states that every exhibit present at a trade show must be constructed to allow for access from persons with disabilities. These specifications can be easily found through numerous resources and online sites. Failing to accommodate for ADA regulations could result in tens of thousands of dollars in fines.

➤ Pay Close Attention to Trade Show Deadlines

Moving in and moving out deadlines need to be honored precisely. Individual booths are given time slots for unloading shipments, and they are given specified times for packing up and moving out after the show is concluded. Clearly communicate all of these deadlines to your shipping team to ensure that those deadlines are met. Failure to do so could cause you to lose your spot in line at the show.

➤ Have Your Exhibit Insured and Have the Exhibitor Insurance Certificate Available

This rule is rather simple to understand. Insurance will cover your booth and its contents in the event of a problem during the show. Show management will need to see proof of your insurance coverage before the show.

➤ Request a "Variance" if you need special accommodations

All of a trade show's rules exist to maintain order among the exhibitors and to help trade show events run smoothly and safely. If you find that you need special accommodations for your booth that do not meet their rules, request a "variance" to have your needs reconsidered. You may get an exemption to have that extra-large banner or specialized music playing at your booth. You won't know unless you ask.

This is not a comprehensive list of exhibition rules and regulations, but it does offer exhibitors a good idea of the types of regulations they can expect. Be sure to check with your trade show manager to discover a comprehensive list of rules to follow. By knowing the rules, and following them closely, you're more likely to have a trade show that goes off without a hitch.

注释:

不当营销推广（Outboarding）目的是利用他人的投资和活动商誉获得好处，不当促销是在酒店客房或招待套房、餐厅、俱乐部或任何其他私人或公共地方进行的商业活动，又称作不当营销活动（Outboarded Event）。该类活动的目的是利用现有展览会主办单位的投资，如活动环节，吸引参会人数，利用展会品牌，吸引参展商参加不当营销活动。活动期间严格禁止此类行为。

不当促销（Suitcasing）是指参会人员在走廊、活动公共区域，或者在其他展位招揽生意的行为。如需向参会人员传播信息和商谈业务，必须购买展馆内的展位空间（如展位或会议室）。所有宣传资料必须从所购买展位内派发，而不得在展馆公共场所或其他参展商展位内派发。

(This article is edited according to Internet information)

NEW WORDS AND PHRASES

hitch   n. 小问题

prospectus   n. 简章, 内容说明书

lure   vt. 吸引, 诱惑

frowned upon   令人发愁

helium balloon   氦气球

duct tape   布基胶带

roam   v. 漫步, 游走

obstruct   vt. 阻碍, 遮挡

time slot   时段

variance   n. 特殊许可

exemption   n. 豁免, 免除

 **Reading 3**

How to Do the Trade Show Logistics the Easy Way

Feeling overwhelmed by the thought of packing and shipping your trade show items to and from the show site? You are not alone. Most first-time exhibitors and even some regular ones have the same concerns.

All the meticulous planning, tons of effort, and sleepless nights that usually go into getting ready for a trade show can be marred if you run into shipping issues. Thankfully, you can avoid this colossal disaster by getting the trade show logistics right. But this doesn't happen by chance. You need to know how to take deliberate steps to ensure smooth trade show execution. Keep reading to learn how.

Trade Show Logistics: A Quick Overview

Trade show logistics involves efficiently moving booth and equipment to a show site, managing setup and utilities, and finally getting everything back to headquarters in one piece after the event.

The entire process can be quite challenging, even if the show is taking place within your city. But it gets tougher when borders are involved. The challenge of managing the logistical side of trade shows in another state or another country is an entirely different ball game. This is where hiring the services of trade show logistics providers can be a big relief.

Whether you choose to do it yourself or outsource to third-party logistics depends on where the show is taking place and the size of your exhibition. Regardless of what works for you, here are some of the ways that trade show logistics can be a game-changer for your business during shows.

Timing

Knowing the time it takes to get from your facility to the site of a show is one thing. Arriving within the window of time allowed by convention centers is another thing altogether. That's because it takes extra hours to load and unload your trade show equipment and there are many usually hundreds of other shipments arriving at the venue at the same time. Ensuring on-time freight shipping delivery is the first step toward a successful trade show outing.

Size and Weight of Equipment

If your exhibition is anything like that of most other trade show participants, it will involve a booth, product displays, and other large pieces of equipment that will probably weigh thousands of pounds. How do you get these heavy items across hundreds of miles? Freight carriers are in a better position to advise on the best form of transport for the size and weight of your shipment. Plus, you can cut down costs by following trade show freight carrier suggestions on fitting all your items into fewer trucks.

Safe and Efficient Delivery

Just about anyone can deliver equipment to a trade show site. However, you're not merely looking to get your booth and items from point A to B; you want a reliable freight carrier that can do so safely and efficiently. This is where a more experienced carrier comes into play. They are less likely to deliver your shipment without damaging them.

Tear Down and Departure

Unlike cargo shipping that involves the hassles of hauling your deliveries back after the show, trade show shipping usually means your equipment ships to headquarters or straight to the next event after the show. This allows you to focus on your business and stick to your schedule.

Tips to Flawless Trade Show Logistics

Trade show logistics are not as difficult as they seem, as long as you have the right approach. To make things a lot easier, the following suggestions will break down the process and make implementation smoother, less overwhelming, and less stressful.

➢ Check the Event's Director before You Get Started

First thing first; reach out to the event's organizers when you register to exhibit at an event. Trade show organizers usually have specific directions regarding equipment. These instructions cover everything you need to know about how to pack trade show items and where to store pre-made booths.

Planning for logistics will be easier if you get this information early, so make sure to ask for the show's logistical direction as early as possible. In most cases, it is more convenient to ship your event equipment well ahead of the show. This means you need to get the schedule of the warehouse and other shipping information beforehand to enable you to check out the details.

Equipped with the event's direction, you will be more confident in planning out your logistics, knowing that everything you need for a successful show will get to the right destination and arrive on time.

➤ Label, Label, Label!

The next step is to label everything. It doesn't matter how large or small the item or container is, label it if you must ship it. If you know anything about trade shows, you'll be quick to agree that event venues are a hive of activity. That's another way of saying trade shows are not the place to send your items to without proper labeling.

Labeling eliminates the chances of getting things mixed up. Of course, you may be able to identify your trade show materials if they are the only ones shipped to the warehouse. But with hundreds or maybe thousands of other exhibits from a large number of companies all stored in the same warehouse, things can get confusing very quickly.

Label every item before storing them in a container and label each container, too. Don't forget to include your brand or company name, event name, and booth number. Remember to add a phone number on boxes, cases, or containers in case of destination mix-up or loss. While you're at it, make sure to label more than one side of your containers or boxes. This will make it easy to identify the box even if items get stacked on top of one of the sides with a label.

Adding an inventory list inside each box is another useful way to painlessly and easily account for your items. Plus, it helps you keep track of what's inside each container when you unpack.

➤ Compare Quotes

It is usually a good practice to get quotes from at least three different logistic providers whether you are shipping items over long or short distances. This is particularly true if it is your first time exhibiting at trade a trade show.

Of course, you can adopt a DIY approach and move your equipment to the show's site yourself, especially if the event happens to be in your city and perhaps you are on a tight budget. It could also be a good option if you have only a few trade show items to convey but it is rather a rare choice. Plus, it may not be your best move if it is your first trade show exhibition. In that case, hiring a company with considerable experience in trade show logistics is your best bet.

Ask for recommendations and don't be shy to work out the best price you can get. Remember to notify your preferred logistic company a few months or several weeks before the event date. This will enable them to fit you into their schedule and also eliminate mistakes that can come from any last-minute rush.

➤ Using the Official Contractor

If you really want to remove unnecessary hassles, simply go with the official contractor or appointed show partner. The official event supplier is already familiar with the landscape and can get you quick results without you worrying so much. Some official contractors can even pick up your e-quipment from your location. However, you need to get in touch with them well in advance and furnish them with all the details of what you'll be moving, and where you want the pick-up.

Keep in mind that event venues and organizers typically don't take deliveries. In other words, choosing to use the official contractor means you need to make arrangements for your staff or a de-livery company to accept delivery of your items from the logistic desk.

➢ Shipping Your Trade Show Materials

Now, we come to one of the most crucial aspects of trade show logistical planning. You have two options when it comes to moving your materials to the site. These are direct to show shipping and advance warehousing.

Advance Warehousing: Ultimately, the choice is yours but we strongly recommend this option if it is available. As earlier mentioned, show venues can be quite chaotic on the day of the event. To minimize confusion and loss of trade show materials, it is common practice for event venues to provide advance warehousing near the show site where exhibitors can store their equipment before the show. Make sure to check the warehousing policies beforehand to know how long you can store your goods. With this option you, don't have to wait for deliveries on the day of the show. You are sure to have access to your booth and materials in good time to allow for early setting up. Advance warehousing is your best bet if you don't want to go through the anxiety of "just-in-time" delivery.

Direct to Show Shipping: With this option, your materials are stored in an off-site location and transported to the show floor on the day of the event. This offers the flexibility of quickly mo-ving your equipment from one event to another with no breaks in between. However, you could runinto serious problems with this option if you are not careful. Picture this scenario: you and a-bout one hundred other companies use the Direct to Show Shipping option to deliver items on the day of the show. It is not hard to figure out the delay that this will cause. The truck conveying your materials could get stuck in a long queue outside the venue. That means you may not be able to set up in time before the opening.

➢ Protect Your Items from Damage While in Transit

There is little you can do than hope that all goes well once you've labeled everything correctly and your trade show items are in transit. However, you can do your part to increase the chances of the items reaching the destination in one piece.

Protecting your items involves simple things you will normally do when traveling with luggage. First, use durable packaging and make sure they are securely sealed. Consider using hard plastic totes or wooden crates if you are moving fragile items.

Wrap breakable items with bubble wraps, padding, or soft cloth. This will protect them from scratches during transit. For insurance, it is usually a good idea to photograph your items before and after wrapping or packing them.

➢ Using Forklifts

You should inform your logistics provider well in advance if you have large items that require the use of forklifts. It is not uncommon for event centers to limit the use of forklifts on the show's site to the official logistic contractor. Make sure to inquire about this beforehand and make the necessary arrangement if you plan to use a forklift.

➤ Who to Talk to If You Have Issues

You can't completely rule out issues no matter how meticulous your trade show logistics plans are. Fortunately, you can talk to a few different people on-site about your challenges. Keep an eye out for signage that says "Hall Manager Office" "Organizers Office" or something similar. These are usually the team of floor managers you can talk to if you find yourself in a jam. You can also identify the organizer team on the floor of the show by their organizer branded uniforms.

➤ Shipping across International Borders

Shipping equipment to an international event is a whole new game ball compared to moving your materials to a domestic trade show. But the real challenge is not merely the distance, although that would cost you more. The main issue is the various regulations and border checks.

International border laws vary from country to country. Brining your trade show goods into one country may be free of charge, while another country may charge you duties. It is common to sign several forms and you may even have tight deadlines. For this reason, shipping across international borders requires a trade show logistics contractor that understands the terrains of international shipping. This will minimize unnecessary delays and get your materials to the site on time.

You must also factor in the long distance and how to meet the time window for delivery. This is particularly important if you have to cross a few different borders to get to the destination country. Ensure that your contractor understands custom requirements to avoid delays that can mar all your planning and preparations.

Undoubtedly, shipping to an international trade show is not an in-house undertaking. You need a specialist company to handle this task, especially one with lots of overseas travel experience. Consider the following reasons for choosing a specialist company if you ever need to ship trade show goods abroad:

Reliable trade show logistics companies who have been in the business of shipping across international borders for a while usually don't have to go through many checks. By contrast, companies providing ad hoc services are subjected to tons of checks and procedures that may lead to annoying delays.

Increased safety and security of materials and booths while in transit from one country to another.

Specialist companies offer insurance for trade show equipment and items. This typically covers the cost of replacing lost or damaged items, including unique and expensive ones.

These companies already have the necessary experience and expertise to avoid pitfalls that can cause delays in delivery. For example, they can get customs clearance for goods ahead of shipment.

➤ Getting Your Items Back Home after the Show

Typically, your logistics supplier will collect your items after the show. But if you've arranged for this, make sure to confirm shipment pick up to avoid forced freight. Make sure you wait until your items are collected before leaving the event venue. Never assume that your equipment is on its way back home, as that might be a costly assumption.

Usually, all items left on the floor of the show after the closing date are considered trash. The organizers will probably dispose of these items and possibly charge you for them. If equipment is held beyond the pullout date, you will be charged for it and the charges will continue to accumulate for each extra day the items are held.

➤ Doing The Trade Show Logistics The Easy Way

Shipping can be stressful, especially if you are new to it. Plus, freight charges are not getting any cheaper. But you don't have to scale back on your exhibition since it is one of the best ways to reach the public. Indeed, the investment might be costly but the returns are can be potentially worth it.

(This article is edited according to Internet information)

NEW WORDS AND PHRASES

meticulous　adj. 一丝不苟的, 小心翼翼的

mar　vt. 损毁

colossal　adj. 巨大的, 异常的

deliberate　adj. 深思熟虑的

hassle　n. 激战

haul　vt. 搬运, 拖运

label　vt. 打标签

hive　n. 热闹的场所

stack　v. 堆放

landscape　n. 地形

chaotic　adj. 混乱的, 无序的

wooden crate　木箱

bubble wrap　泡沫包装

forklift　n. 升降机

terrain　n. 领域

pitfall　n. 陷阱, 圈套

pullout　n. 撤离

scale back　缩减

ENGLISH FOR WORKPLACE COMMUNICATION

Sample Dialogue: Booking a booth.

Situation: The following conversation is between Cathy (C), a staff from an exhibition company, and Zhong (Z), sales manager from ABC company. Zhong wants to book a perimeter booth, but it is the first time that his company participates in the exhibition.

C: Good morning.

Z: Good morning, I am Zhong, the sales manager from ABC company.

C: How may I help you?

Z: I heard that an exhibition will be held in April. could you tell me more details about it?

C: Sure. We have island booth, peninsula booth, perimeter booth and linear booth. And this is our catalogue.

Z: Thanks. I'd like to book a perimeter booth, about normal size, but I think the price is too expensive.

C: Is this the first time you participate in our exhibition?

Z: Yes!

C: We have a 5% discount for new exhibitor.

Z: Oh! I see. That would be better than before.

C: And here are some forms you need to fill in.

Z: Thank you. when should I submit?

C: Before January 30th.

Z: As you know. We are the first time to participate in the exhibition, what should we pay attention to?

C: You should attract your visitors by your booth design.

Z: Sure. Can I arrange some people to hold a billboard, and walking in the exhibition?

C: It is also a good idea.

Z: I see. Thanks for your support.

C: Our pleasure. Thanks for your coming. Bye!

Z: Bye!

TRADE FAIR CITIES AND HUMAN GEOGRAPHY

通过本章的学习，了解全球主要的会展城市和这些城市所在国家的人文地理。了解一个国家、一个城市的经济、文化、风俗、生活习惯等对于企业通过展览会开拓国际市场有着积极的作用。

本章重点介绍了德国法兰克福、美国拉斯维加斯和阿联酋迪拜，欧洲、美洲和中东地区有着迥然不同的风俗和文化。从天主教、基督教到伊斯兰教，从德国人的严谨到美国人的热情，再到阿联酋人的各种清规戒律，我们畅游了三个国家和三个城市。

积累人文地理知识是一个长期的过程，读书、旅行、工作与聊天都是获取知识和信息的重要途径，而从事会展工作是增长人文地理知识最快捷的通道之一。

知识精讲

展会知识

2020 年迪拜世博会后记，向未来转型？

在用影像记录 2020 年迪拜世博会的过去两年中，我时常被朋友问到它究竟是什么样的。为了配合它在 3 月 31 日的闭幕，我想要从两个角度解答这个问题。大部分读者都知道，世博会可追溯到 19 世纪，是企业和国家的展示柜，是旅游景点，又是技术创新的试验场。但它同时也是对于机会、可持续性和流动性的范例研究集合，以及一个有利可图的建筑项目，在阿联酋创造一个新的社区以及未来的居住模式。或许最后这一参照系是最有趣的，因为它是这次世博会与其他世博会的最大区别之处。

一点背景

考虑到在过去的 171 年里，世博会每五年就会在世界各地举办一次，人们大部分都会对其有一知半解。1851 年第一届世博会在伦敦举办，目的是推动工业革命的技术创新。从那时起，规模已经不断壮大。在 2020 年世博会上，有 192 个国家参展，来自世界各地的 20 万工作人员聚集在一起，使这一梦想变为现实。来自 179 个国家的超过 3 万家企业在此注册开展业务。

迪拜的旅游业让巴塞罗那和威尼斯等城市望尘莫及。2019 年，在六个月内大约有 1 000 万人前往威尼斯旅游，700 万前往巴塞罗那。据 Travel Euromonitor International 资深分析师 Rabia Yasmeen 称，迪拜在 2018 年已经成为世界上第七大旅游城市，年均游客超过 1 600 万。此外，迪拜也是游客消费最多的城市，2018 年的消费总额高达 3 000 万迪拉姆。相比之下，当年巴黎的游客消费仅为 1 400 万迪拉姆。直至今日，仅在六个月内，2020 世博会已经拥有了 19 009 065 位游客以及 1.8 亿在线游客，使其比兰布拉斯更像一个观光胜地。

这些活动一直都是一种技术奥林匹克盛会。历届世博会上展示的惊人发明包括：X 光机器、打字机、第一次电视直播、IMAX 电影、亨氏番茄酱以及蛋卷冰激凌等。本届世博会也不例外，每个展厅都拥有丰富的小玩意。创新影响资助计划（IIGP）是实现这一目标的部分原因，来自 76 个国家的 140 名受资助者获得了资金、指导和曝光。已经收到了来自 184 个国家（约占世界上所有国家的 96%）的超过 11 000 份申请。除此之外，世博会还提供了大学创新计划，为阿联酋的 46 组学生提供了每组 25 000~50 000 迪拉姆的奖励。

一个有利可图的建筑项目在阿联酋创造了一个新的社区

所有一切都是极佳的，但建筑创新可能是其中最重要的。规划师和设计师肩负着世博会遗产的责任，保证为期六个月的大会结束之后，80% 的建筑还能继续使用。该项目将保留 2020 年迪拜世博会的总建筑面积超过 26 万平方米的 LEED 金奖和白金建筑。这些将会被分阶段改造成为 2020 区域的住宅、商业以及文化街区。由 Woods Bagot 设计的迪拜展览中心（DEC）提供了 45 000 平方米的定制空间，坐落于 60 000 平方米的场地内，拥有 14 个多功能大厅、4 个套间以及 24 个会议室。该建筑过于巨大，我发现将其装入我最宽的镜头都是种挑战。

4.38 平方千米的世博园区被划分为三个主要区域：流动性、机遇以及可持续性，由 Al Wasl 广场连接在一起。每个区都拥有一个主题展馆。

霍普金斯建筑事务所（Hopkins Architects）的首席建筑师、总规划师和主题区首席设计师 Simon Fraser 这样描述他的设计动力：“我一直认为所有的世博会都与三个重要方面相关，即在建筑和场所中可以表达什么样的特点和语境含义，这些在大会之后可以如何将其重新利用，最后也最重要的是，一个人在哪里可以找到一个休息的地方，在闲逛这些展馆之后从‘博物馆疲乏’中得到缓解。”

我很幸运，得到了委托拍摄霍普金斯的这个项目，而正是这个任务让我一直坚持回来。因此，我对于世博会的许多感受都来源于 Simon——它的样子，它的意义。因此让我们继续他的话题：“我认为我们的总体规划和建筑需要有阿联酋的特点，以及一种非常灵活、适应性强的松散的建筑语言，这是合适的。回望迪拜小河附近的 Bastakia 老城区，漫步在狭窄的街道和景观庭院中的确是一种享受，它们拥有极佳的人体尺度和特征，使用一致但重复的材料和细节，却都有略微不同的表达方式。”

可持续发展园区的另一个重要特征是 Terra，一间由 Grimshaw 建筑事务所设计的可持续展馆。展馆的顶棚可容纳超过 6 000 平方米的嵌入玻璃面板的超高效单晶光伏电池。电池和玻璃外壳的结合使建筑在获得太阳能的同时也为下方的游客提供遮阳和采光。庭院的体验是身处一个巨大的树荫之下，斑驳的光线投射到下面的表面。顶棚的形式与庭院配合，引导冷空气进入，同时通过中心的烟囱效应排出低洼的热空气。顶棚同时也作为雨水和露珠的大型收集区域，补充建筑的水系统。

我被委任拍摄这座场馆以及下面一系列图片中的一些国家展馆，希望能够让人们更广泛地了解迪拜世博会是什么样子。新加坡、荷兰以及芬兰是我的最爱。

担忧

在询问世博会到底是什么样的之后，与我交谈的人们反复提到一系列担忧。可持续性是其中之一。也许对于能源来说，最可持续的利用方式是什么都不建。另一个选择是为一个像这样的大会进行建造，这样它将在活动本身之外持续多年。再次引用霍普金斯的 Simon 的话："我也认为你能做的最可持续的事情之一是让建筑具有灵活性和适应性——我们的想法集中在这一点上，这样所有展馆都可以以最少的改造过程来改变。"最佳实践对于这种规模的项目至关重要。除了遗产建筑，展馆和园区都满足了 LEED 黄金标准，这是能源与环境设计领导力体系四个级别中的第二高标准。如前所述，许多建筑已经成为可持续建筑方式的研究平台。而且，世博会为迪拜地铁的大规模扩建提供了机会。世博会开幕后，每天都乘坐地铁前往场地的我可以证明它的受欢迎程度（有时几乎像沙丁鱼罐头一样）以及使用时的舒适和方便。我发现，在所有尝试过的方式之中，地铁是目前为止往返迪拜世博会的最佳方式。

另一个共同关注的话题是人权。在世博会建设期间，我曾多次见证阿联酋的情况。可以说，入职培训很彻底，执行明确（我为了在不同场合获准进入现场，分别经历了几次培训），项目的组织工作是我以前从未见过的。工人团队被分为几百组，每个小组都拥有颜色标记的安全帽，在大型的尘土飞扬的复杂建筑工地中移动，说着几十种语言，反映了这次行动的组织天才。我带着相机和三脚架走在这些勤劳的人群之中时，确实感到心生敬畏。不仅仅是因为操作的复杂性，也因为我有一半的时间在高温下处于晕倒的边缘，而且工作也不像他们那么辛苦。幸运的是，没有人在一天之中最热的时间工作，我去过的所有地方都有水供应。虽然如此，这也真的是极其艰苦的工作。建造人员与高温、灰尘、噪声、压力抗争多年，以实现这个项目。所有建筑场地都是一种利益冲突与后勤噩梦的战场。但这也同时是我见过的规模最大也最有组织的战役。这是一个令人愉悦的地方吗？不。它是残酷困难的。但它甚至本可以更糟，这要归功于那些身处金属色总部（令人想起迪拜的警察局）长期受苦的组织者以及汗流浃背、臭气熏天、满身灰尘、坚忍不拔的建造者。

平等在整个行业都是一个议题，但在阿联酋似乎并不比在西方严重。同样的，我分享的只是一些个人印象，但在参观了那里大多数建筑工作室之后，我能证明每个工作室都有许多女性建筑师在工作。她们的着装从西式到传统阿联酋式的不等，因为阿联酋是个十字路口和大熔炉，许多其他类型的服装也很受欢迎。这些办公室里的通信人员大多是女性，大部分领导者是男性，这并不是该地区特有的现象。只有墨西哥城在我拍摄建筑工作室的经历中独树一帜，有几间由女性拥有并运营的工作室。

至于"连接思想，创造未来"，这没有什么可说的。公关口号总是让人想起那些空洞、不真诚的商业术语——这是人类还是机器人所写的？它们就像公共艺术，不冒犯也满足不了任何人。它们和那些建筑摄影师被要求拍摄的蓝天照片一样。清淡、易于消化，明确。毫无疑问，为了让这次大会成功举办，人们的思想连接了起来，而且在启动时也是相通的。建筑总是在某种意义上创造未来。或许这就是它是动名词的原因，现状永远在向前进。为了表现它们对于这项遗产很感兴趣，作为如何让事情更好地向前发展的研究，以及认真对待从石油到可持续能源的转变，为何不能将其称为"向未来转变"并将其与一系列承诺相联系起来？但标语并不是重点。重点是，它有什么好处？以什么标准？我对大多数事物都有适应性重用的偏好，无论是历史的，还是现代的但符合当地环境的成语，我都不喜欢这样。这是一个艰难的推销。但在过去两年参观迪拜世博会期间，我的敬意随着它所展现的纯粹的成就而增长。而且有些建筑相当可爱。主题园区的确是老城的现代版本——在许多意义上是更好的一个。遮阳结构对于光影的把握，使用水体以提供凉爽和舒缓，加上鸟鸣（它们喜欢那里），真是美极了。它简直就是沙漠中的绿洲。

Terra 和 Alif 令人着迷。希望它们可以鼓舞建造其他许多拥有类似精神的建筑物。我对它们爱不释手，并对每一个壮举感到敬畏，它们是多么地与目的相配，建造得多么精良，它如何随着光影变化。也许有一天，国家馆将效仿密斯凡德罗的开创性展馆（除了现代馆以外的所有建筑）的优雅。当今的趋势是"向拉斯维加斯学习"——看着一些内部只有迷人的屏幕确实让人感到心痛。但他们确实在拉斯维加斯拥有大量游客。根据这个标准，人们将如何从装饰过的棚子里搬走呢？

（作者 Marc Goodwin，Archmospheres｜译者 July Shao）

思考并回答以下问题：

1. 世博会是怎样的一场展览会，与传统意义上的商品展有什么区别？

2. 迪拜世博会的主题是什么？通过文章中的描述，迪拜世博会是如何实现这一主题的？

3. 迪拜世博会受到全球疫情的影响，推迟了一年举办，但名字仍然使用 2020 年迪拜世博会，日本的奥运会也是一样，推迟一年，也是使用了 2020 年东京奥运会的名字。有一种说法，世博会就是展览行业的奥运会，你是如何理解的？

4. 分组讨论世博会的定义、意义、历史、申办程序。

5. 上海世博会是哪年举办的？主题是什么？对中国的经济发展带来了什么样的影响？

📖 Reading 1

Frankfurt am. Main

Frankfurt, officially Frankfurt am Main, is the most populous city in the German state of Hesse. Its 763,380 inhabitants as of 31 December 2019 make it the fifth-most populous city in Germany. On the river Main (a tributary of the Rhine), it forms a continuous conurbation with the neighboring city of Offenbach am Main and its urban area has a population of over 2.3 million. The city is the heart of the larger Rhine-Main metropolitan region, which has a population of more than 5.6 million and is Germany's second-largest metropolitan region after the Rhine-Ruhr region. Frankfurt's central business district, the Bankenviertel, lies about 90 km northwest of the geographic center of the EU at Gadheim, Lower Franconia. Like France and Franconia, the city is named after the Franks. Frankfurt is the largest city in the Rhine Franconian dialect area.

Frankfurt was a city state, the Free City of Frankfurt, for nearly five centuries, and was one of the most important cities of the Holy Roman Empire, as a site of Imperial coronations; it lost its sovereignty upon the collapse of the empire in 1806, regained it in 1815 and then lost it again in 1866, when it was annexed (though neutral) by the Kingdom of Prussia. It has been part of the state of Hesse since 1945. Frankfurt is culturally, ethnically and religiously diverse, with half of its population, and a majority of its young people, having a migrant background. A quarter of the population consists of foreign nationals, including many expatriates. In 2015, Frankfurt was home to 1909 ultra-high-net-worth individuals, the sixth-highest number of any city.

Frankfurt is a global hub for commerce, culture, education, tourism and transportation, and rated as an "alpha world city" according to GaWC. It is the site of many global and European corporate headquarters. In addition, Frankfurt Airport is the busiest in Germany, one of the busiest in both Europe and the world, the airport with the most direct routes in the world, and the primary hub for Lufthansa, the national airline of Germany.

Frankfurt is one of the major financial centers of the European continent, with the headquarters of the European Central Bank, Deutsche Bundesbank, Frankfurt Stock Exchange, Deutsche Bank, DZ Bank, Commerzbank, several cloud and fintech startups and other institutes. Automotive, technology and research, services, consulting, media and creative industries complement the economic base. Frankfurt's DE-CIX is the world's largest internet exchange point.

Messe Frankfurt sets up business model for global business. Messe Frankfurt is one of the world's leading trade fair, congress and event organizers with their own exhibition grounds. For some 800 years, we have been bringing people together at our events, both in Frankfurt and throughout the world. From automotive to logistics, from textiles to music, from energy efficiency to security, from homes to beauty-Messe Frankfurt's programme of trade fairs, congresses and other events is extremely comprehensive. With our events we meet highest quality standards and create global interfaces between the industry, commerce, politics, services and consumer goods sectors. Whether with our trade fairs, congresses or other events, with digital or analogue services-everything we do has always been focused on interaction between people. This is something that we have perfected throughout our long history.

With approximately 2,200 employees at 28 locations, we work virtually around the clock and around the globe to further the interests of our customers. Fairs & Events is where we bundle our international interaction formats. Despite the very difficult conditions in 2021, some 1.4 million visitors (2019: 5.1 million) and more than 30,000 (2019: 99.246) exhibiting companies put their trust in the efficiency of our international network the quality of our events and the digital expertise of Messe Frankfurt. In 2021, 187 trade fairs, congresses and events around the globe were organized under the umbrella of Messe Frankfurt both physically and digitally (2019: 423), including 64 trade fairs or exhibitions (2019: 155). We know exactly which future trends are of burning importance to our customers and have close ties with business and policymakers. We are expanding our expertise in defined industry sectors and helping our customers achieve their growth targets with specialized innovation forums.

We act as an interface between supply and demand and between trends and markets. Messe Frankfurt offers tailor-made trade fair formats worldwide with an international target audience. As a reliable marketing partner, we are synonymous with performance, quality and trust. This can be seen from the success of our customers. We act as a marketing partner to our customers in the following industry sectors:

Consumer Goods: Both as everyday utensils and design objects, they are an important economic factor. They can also contribute pure emotion to our lives. Globalization, digitization and new distribution forms are changing the industry at a breakneck pace. With a brand family that includes Ambiente and Beautyworld, Messe Frankfurt has the internationally leading business platforms that help business take advantage of the opportunities of the future. With a series of leading consumer goods trade fairs focusing on design, tableware, gifts and home décor, Messe Frankfurt offers unique marketing and sales opportunities. Ambiente worldwide enables exhibitors and trade visitors to make new business contacts and conclude major transactions at its shows not only in Frankfurt, but also in Tokyo, Dubai, Mumbai and Shanghai. In this sector we also have Paperworld, Christmasworld, Beautyworld, Tendence Lifestyle.

Building Technologies: Digitization, networking, safety and security are some of the most pressing issues of our time. Messe Frankfurt has 26 events in Europe, Asia and South America that address these themes and offer key solutions. They provide innovative interaction platforms for efficient building management, intelligent networking, security of supply, design and sustainable usage of scarce resources, namely energy and water. The two international flagship events ISH and Light + Building are of particular interest here.

Safety, Security & Fire: Messe Frankfurt is one of the world's leading organizers of events for the growth field of civil security, with eleven events in Asia, Germany, the Middle East, Russia, South America and East Africa. We have responded to the growing global market in this sector by offering an excellent means of gaining access to the world's most dynamic regions with international brands such as Intersec and Secutech.

Environmental Technologies: The increasing scarcity of fossil fuels, growing global energy demand, access to clean water, waste disposal and recycling are all amongst the major challenges facing the world in future. These challenges are also creating new opportunities in these markets worldwide. The world needs intelligent technologies and innovative companies. Eco Expo Asia is an example of the platforms we offer throughout the world to foster exchange in the growing market for environmental technology.

Food Technologies: Sustainable and responsible food management for a growing global population presents challenges for which the industry has concrete solutions. Innovative ways of ensuring the safe, secure, efficient and economical production and packaging of food are the focal point of our four trade fairs for the food industry. The industry's top themes are resource efficiency, production optimization, food safety, digital solutions and food trends. With IFFA, an event that has been taking place since 1949, Messe Frankfurt is responsible for organizing the international flagship event for the meat industry in Frankfurt am Main. The company's other three events in this industry are located in China and Argentina.

Textiles & Textile Technologies: Messe Frankfurt created the "Texpertise Network" to bring together its expertise as global market leader for organizing trade fairs and other events for the textile industry, including leading international trade fairs such as Heimtextil, Techtextil, Texprocess, Texworld and Intertextile. We showcase the trends and themes that are driving the textile and fashion industries at about 60 trade fairs worldwide with more than 23,000 exhibiting companies and over 600,000 trade visitors. Be it Berlin, Buenos Aires, Frankfurt, New Delhi, New York,

Paris, Shanghai, Tokyo or Addis Ababa—we create momentum for the entire textile value chain.

Mobility & Logistics: The digitization and decarbonization megatrends are having a dramatic impact on mobility and logistics. Connected cars, autonomous driving and renewable energy drive systems are the hot-button topics in the sector. At our international events such as Automechanika and Hypermotion, it is not only key accounts from the automotive sector who engage in dialogue. We provide the right interaction formats for intelligent ideas relating to transport systems of the future. Not to mention events in the fields of commercial vehicles, motorcycles and logistics.

Manufacturing Technologies & Components: Intelligent process optimization, complexity and agility: Messe Frankfurt provides focused B-to-B platforms for each of these fields in growth regions worldwide. These include trade fairs and congresses, such as Formnext for additive manufacturing and intelligent production processes as well as Wire & Cable Guangzhou for the wire and cable industries.

Textile Care, Cleaning & Cleanroom Technologies: With the ever-increasing importance of Industry 4.0, digitization, safety, security, textile care and cleanrooms have become essential components. Messe Frankfurt events throw a spotlight on industry's latest innovations in key fields, creating the ideal platform for discussions with business partners and customers. With Texcare International in Frankfurt, Texcare Asia in Shanghai and various conferences taking place under the Texcare Forum brand, Messe Frankfurt has established a presence in strategic markets for textile care, while the specialist Cleanzone fair focuses on hygiene and cleanroom technologies.

Event & Entertainment Technologies: Shows and other forms of live performances are taken to a new level of experience. This is made possible by technology and services relating to production, staging and digital networking. The international Prolight + Sound brand showcases the latest industry trends. This is where decision-makers, planners, creatives and visionaries get together with current and future professionals.

Messe Frankfurt brings together future trends with new technologies, people with markets, and supply with demand. Where different perspectives and industry sectors come together, we create scope for new collaborations, projects and business models.

Römer, the German word for Roman, is a complex of nine houses that form the Frankfurt city hall (Rathaus). The houses were acquired by the city council in 1405 from a wealthy merchant family. The middle house became the city hall and was later connected with its neighbors. The Kaisersaal ("Emperor's Hall") is located on the upper floor and is where the newly crowned emperors held their banquets. The Römer was partially destroyed in World War II and later rebuilt. The surrounding square, the Römerberg, is named after the city hall.

Frankfurt Central Station (Frankfurt Hauptbahnhof), which opened in 1888, was built as the central train station for Frankfurt to replace three smaller train stations in the city center and to boost the needed capacity for travelers. It was constructed as a terminus station and was the largest train station in Europe by floor area until 1915 when Leipzig Central Station was opened. Its three main halls were constructed in a neorenaissance-style, while the later enlargement with two outer halls in 1924 was constructed in neoclassic-style.

Zeil-Frankfurt's central shopping street. It is a pedestrian-only area and is bordered by two large public squares, Hauptwache in the west and Konstablerwache in the east. It is the second most expensive street for shops to rent in Germany after the Kaufingerstraße in Munich. 85 percent of the shops are retail chains such as H&M, Saturn, Esprit, Zara or NewYorker. In 2009 a new shopping mall named MyZeil opened there with nearly 100 stores and chains like Hollister. Three more shopping malls occupy the Zeil: Galeria Kaufhof and Karstadt, as well as large fashion retail clothing stores from Peek & Cloppenburg and C&A. During the month before Christmas, the extended pedestrian-only zone is host to Frankfurt Christmas Market, one of the largest and oldest Christmas markets in Germany.

With a large forest, many parks, the Main riverbanks and the two botanical gardens, Frankfurt is considered a "green city": More than 50 percent of the area within the city limits are protected green areas. With more than 30 museums, Frankfurt has one of the largest variety of museums in Europe. Twenty museums are part of the Museumsufer, located on the front row of both sides of the Main riverbank or nearby, which was created on an initiative by cultural politician Hilmar Hoffmann. Ten museums are located on the southern riverbank in Sachsenhausen between the Eiserner Steg and the Friedensbrücke. The street itself, Schaumainkai, is partially closed to traffic on Saturdays for Frankfurt's largest flea market.

Oper Frankfurt—A leading Germany opera company and one of Europe's most important. It was elected Opera house of the year (of Germany, Austria and German-speaking Switzerland) by German magazine Opernwelt in 1995, 1996 and 2003. It was also electedthe best opera house in Germany in 2010 and 2011. Its orchestra was voted Orchestra of the year in 2009, 2010 and 2011.

Frankfurter Würstchen—"short Frankfurter" is a small sausage made of smoked pork. They are similar to hot dogs. The name Frankfurter Würstchen has been trademarked since 1860. Frankfurter Rindswurst is a kind of sausage made of pure beef. Frankfurter Rippchen-Also known as Rippchen mit Kraut, is a traditional dish which consists of cured pork cutlets, slowly heated in sauerkraut or meat broth, and usually served with sauerkraut, mashed potatoes and yellow mustard. Crispy pork knuckle with saute potatoes (Eisbein), is a traditional German dish of juicy, tender pork knuckle with the crispiest, crackly skin. Serve with vegetables and perfectly cooked creamy sauté potatoes for a meal you will want to eat again and again!

The city can be accessed from around the world via Frankfurt Airport (Flughafen Frankfurt am Main) located 12 km (7 mi) southwest of the city center. The airport has four runways and serves 265 non-stop destinations. Run by transport company Fraport it ranks among the world's busiest airports by passenger traffic and is the busiest airport by cargo traffic in Europe. The airport also serves as a hub for Condor and as the main hub for German flag carrier Lufthansa. It is the busiest airport in Europe in terms of cargo traffic, and the fourth busiest in Europe in terms of passenger traffic behind London Heathrow Airport, Paris Charles de Gaulle Airport and Amsterdam Airport Schiphol. Passenger traffic at Frankfurt Airport in 2018 was 69, 510, 269 passengers. A third terminal is being constructed (planned to open in 2023). The third terminal will increase the capacity of the airport to over 90 million passengers per year.

The airport can be reached by car or bus and has two railway stations, one for regional and one for long-distance traffic. The S-Bahn lines S8 and S9 (direction Offenbach Ost or Hanau

Hbf) departing at the regional station take 10-15 minutes from the airport to Frankfurt Central Station and onwards to the city center (Hauptwache station), the IC and ICE trains departing at the long-distance station take 10 minutes to Frankfurt Central Station.

The Frankfurt Trade Fair offers two railway stations: Messe station is for local S-Bahn trains (lines S3-S6) and is located at the center of the trade fair premises while Festhalle/Messe station is served by U-Bahn line U4 and is located at the north-east corner of the premises.

Frankfurt is one of the world's most important financial centers and Germany's financial capital, followed by Hamburg and Stuttgart. Frankfurt is home to two important central banks: the German Bundesbank and the European Central Bank (ECB). The European Central Bank (Europäische Zentralbank) is one of the world's most important central banks. The ECB sets monetary policy for the Eurozone, consisting of 19 EU member states that have adopted the Euro (€) as their common currency.

The Frankfurt Stock Exchange (Frankfurter Wertpapierbörse) began in the ninth century. By the 16th century Frankfurt had developed into an important European hub for trade fairs and financial services. Today the Frankfurt Stock Exchange is by far the largest in Germany, with a turnover of more than 90 percent of the German stock market and is the third-largest in Europe after the London Stock Exchange and the European branch of the NYSE Euronext. The most important stock market index is the DAX, the index of the 30 largest German business companies listed at the stock exchange.

(This article is edited according to Internet information)

NEW WORDS AND PHRASES

Frankfurt am Main　法兰克福(美因河畔)

inhabitant　n. 居民

tributary　n. 河流的分支

conurbation　n. 邻近城市因扩建而形成的组合城市

metropolitan　adj. 大都市的

Franconian dialect area　法兰克尼亚方言区

Imperial coronation　帝王加冕

sovereignty　n. 主权

collapse　v. 瓦解,崩塌

annexed　adj. 附属的

ethnically　adv. 种族上

expatriate　n. 外派人员

hub　n. 中心

Lufthansa　n. 德国汉莎航空公司

fintech　n. 金融科技

Messe Frankfurt　法兰克福展览有限公司

innovation forum　创新论坛

tailor-made　adj. 定制的,特制的

synonymous   adj. 同义的, 紧密联系的

utensil   n. 用具

breakneck   adj. 极快的

flagship   n. 一流

dynamic   adj. 充满活力的

scarcity   n. 不足, 缺乏

foster   v. 促进

concrete solution   具体的解决方案

focal   adj. 焦点的

momentum   n. 动力

decarbonization megatrend   脱碳大趋势

hot-button   n. 热点问题

agility   n. 敏捷, 灵活

terminus   n. 终点

pedestrian   adj. 徒步的

botanical garden   植物园

orchestra   n. 管弦乐队

turnover   n. 营业额

 EXERCISES

Ⅰ. Answer the following questions.

1. According to this article, what factors are changing the consumer goods industry at a breakneck pace?

2. What are the most pressing issues of our time according to the article?

3. What are the major challenges facing the world in future?

4. What are the top themes in Food Technologies industry?

5. What are the hot-button topics in the mobility and logistics sector?

6. What have become essential with the ever-increasing importance of Industry 4.0?

Ⅱ. Please translate the following English sentences into Chinese.

1. With our events we meet highest quality standards and create global interfaces between the industry, commerce, politics, services and consumer goods sectors.

2. With approximately 2,200 employees at 28 locations, we work virtually around the clock and around the globe to further the interests of our customers.

3. We act as an interface between supply and demand and between trends and markets.

4. As a reliable marketing partner, we are synonymous with performance, quality and trust.

5. These challenges are also creating new opportunities in these markets worldwide.

6. The industry's top themes are resource efficiency, production optimization, food safety, digital solutions and food trends.

7. We create momentum for the entire textile value chain.

8. Messe Frankfurt brings together future trends with new technologies, people with markets,

and supply with demand.

9. With more than 30 museums, Frankfurt has one of the largest variety of museums in Europe.

10. Frankfurt is one of the major financial centers of the European continent.

Ⅲ. There are 10 sentences in this section. Beneath each sentence there are four words or phrases marked A, B, C and D. Choose one word or phrase that best completes the sentence.

1. Frankfurt is the (　　) -most populous city in Germany.

A. second　　　　　B. third　　　　　C. fourth　　　　　D. fifth

2. Where is the busiest airport located in Germany? (　　)

A. Berlin　　　　　B. Hamburg　　　　C. München　　　　D. Frankfurt

3. Where is the headquarters of the European Central Bank located? (　　)

A. London　　　　　B. Paris　　　　　C. Rome　　　　　D. Frankfurt

4. How manytrade fairs or exhibitions were organized under the umbrella of Messe Frankfurt in 2021? (　　)

A. 187　　　　　　B. 155　　　　　　C. 64　　　　　　D. 423

5. Which one is not the business platform related to the consumer goods industry? (　　)

A. Beautyworld　　B. Ambiente　　　　C. Paperworld　　　D. Intersec

6. ISH is aninternational flagship event for (　　).

A. Safety, Security & Fire　　　　　B. Building Technologies

C. Environmental Technologies　　　D. Consumer Goods

7. Eco Expo Asia is a platform for (　　).

A. Building Technologies　　　　　B. Environmental Technologies

C. Food Technologies　　　　　　D. Mobility & Logistics

8. The Texcare Asia is in (　　).

A. Tokyo　　　　　B. Hong Kong　　　C. Shanghai　　　　D. Beijing

9. "Short Frankfurter" are much similar to (　　).

A. hamburger　　　B. hot dogs　　　　C. pizza　　　　　D. spaghetti

10. (　　) is the largest Stock Exchange in Europe.

A. Frankfurt Stock Exchange

B. London Stock Exchange

C. European branch of the NYSE Euronext

D. Amsterdam Stock Exchange

Ⅳ. Writing

Please write an inquiry with the following particulars:

1. You are the general manager of the United Textiles Co, Ltd, and now you need to write an inquiry to Lily.

2. You and Lily (the sales manager of China ABC Textiles Company) met with each other at a textile trade fair in Frankfurt.

3. You are interested in silk products and hope to receive best quotation CIF London for 2000

pieces of silk carpet.

4. You need the details of discount and payment terms. Catalogue and sample cuttings are also required.

 **Reading 2**

Las Vegas

Las Vegas (Spanish for "The Meadows"), often known simply as Vegas, is the 25th-most populous city in the United States, the most populous city in the state of Nevada, and the county seat of Clark County. The city anchors the Las Vegas Valley metropolitan area and is the largest city within the greater Mojave Desert. Las Vegas is an internationally renowned major resort city, known primarily for its gambling, shopping, fine dining, entertainment, and nightlife. The Las Vegas Valley as a whole serves as the leading financial, commercial, and cultural center for Nevada.

The city bills itself as The Entertainment Capital of the World, and is famous for its luxurious and extremely large casino-hotels together with their associated activities. It is a top three destinationin the United States for business conventions and a global leader in the hospitality industry, claiming more AAA Five Diamond hotels than any other city in the world. Today, Las Vegas annually ranks as one of the world's most visited tourist destinations. The city's tolerance for numerous forms of adult entertainment earned it the title of "Sin City", and has made Las Vegas a popular setting for literature, films, television programs, and music videos.

Las Vegas was settled in 1905 and officially incorporated in 1911. At the close of the 20th century, it was the most populated North American city founded within that century (a similar distinction was earned by Chicago in the 19th century). Population growth has accelerated since the 1960s, and between 1990 and 2000 the population nearly doubled, increasing by 85.2%. Rapid growth has continued into the 21st century, and according to the United States Census Bureau, the city had 641,903 residents in 2020, with a metropolitan population of 2,227,053.

As with most major metropolitan areas, the name of the primary city ("Las Vegas" in this case) is often used to describe areas beyond official city limits. In the case of Las Vegas, this especially applies to the areas on and near the Las Vegas Strip, which are actually located within the unincorporated communities of Paradise and Winchester. Nevada is the driest state, and Las Vegas is the driest major U. S. city. Over time and influenced by climate change, droughts in Southern Nevada have been increasing in frequency and severity, putting a further strain on Las Vegas' water security.

Las Vegas has a subtropical hot desert climate, typical of the Mojave Desert in which it lies. This climate is typified by long, extremely hot summers; warm transitional seasons; and short winters with mild days and cool nights. There is abundant sunshine throughout the year, with an average of 310 sunny days and bright sunshine during 86% of all daylight hours. Rainfall is scarce, with an average of 4.2 in (110 mm) dispersed between roughly 26 total rainy days per year. Las Vegas is among the sunniest, driest, and least humid locations in North America, with exceptionally low dew points and humidity that sometimes remains below 10%.

The primary drivers of the Las Vegas economy are tourism, gaming, and conventions, which in turn feed the retail and restaurant industries.

The major attractions in Las Vegas are the casinos and the hotels, although in recent years other new attractions have begun to emerge. Most casinos in the downtown area are located on Fremont Street, with The STRAT Hotel, Casino & Skypod as one of the few exceptions. Fremont East, adjacent to the Fremont Street Experience, was granted variances to allow bars to be closer together, similar to the Gaslamp Quarter of San Diego, the goal being to attract a different demographic than the Strip attracts.

The city is home to an extensive Downtown Arts District, which hosts numerous galleries and events including the annual Las Vegas Film Festival. "First Friday" is a monthly celebration that includes arts, music, special presentations and food in a section of the city's downtown region called 18b, The Las Vegas Arts District. The festival extends into the Fremont East Entertainment District as well. The Thursday evening prior to First Friday is known in the arts district as "Preview Thursday", which highlights new gallery exhibitions throughout the district.

Las Vegas has 68 parks. The city owns the land for, but does not operate, four golf courses: Angel Park Golf Club, Desert Pines Golf Club, Durango Hills Golf Club, and the Las Vegas Municipal Golf Course. It is also responsible for 123 playgrounds, 23 softball fields, 10 football fields, 44 soccer fields, 10 dog parks, six community centers, four senior centers, 109 skate parks, and six swimming pools.

RTC Transit is a public transportation system providing bus service throughout Las Vegas, Henderson, North Las Vegas and other areas of the valley. Inter-city bus service to and from Las Vegas is provided by Greyhound, Bolt Bus, Orange Belt Stages, Tufesa, and several smaller carriers. Amtrak trains have not served Las Vegas since the service via the Desert Wind at Las Vegas station ceased in 1997, but Amtrak California operates Thruway Motorcoach dedicated service between the city and its passenger rail stations in Bakersfield, California, as well as Los Angeles Union Station via Barstow. The Las Vegas Monorail on the Strip was privately built, and upon bankruptcy taken over by the Las Vegas Convention and Visitors Authority.

Las Vegas is one of the top trade show destinations in the world, and there are many reasons that so many people are setting up their trade show exhibits in Las Vegas. This city offers so many benefits, from providing great accommodation options to offering a stunning cityscape that everyone enjoys seeing. Are you trying to decide where to have your next big trade show or convention? Here's a closer look at 10 of the best reasons to hold your next trade show in Vegas.

➢ Plenty of Trade Show and Convention Space in the City

Las Vegas is filled with plenty of trade show and convention space, making it possible to set up trade shows of any size. With a little investigation, you can find a great space for your show.

➢ Thousands of Hotel Rooms Available

The city boasts more than 150,000 hotel rooms, which means there are plenty of accommodations for trade show attendees. The selection is huge, and there's a spot in Vegas that fits everyone's budget, which is another reason it's a hot spot for trade shows.

➤ The International Airport is Close By

McCarran International Airport is Just a few miles away from Downtown Las Vegas. This makes it easy for people from all over the world to come in for trade show display and exhibits in Las Vegas.

➤ Plenty of Transportation Options

Transportation is always important when choosing a city for a trade show, and Las Vegas offers plenty of options. The Las Vegas Monorail offers a great transportation option, and plenty of limousine and taxi services are also available to help get visitors to exhibits in Las Vegas.

➤ Sunny, Predictable Weather

The weather in Vegas is nearly always sunny, and it's very predictable. It's easy to predict theseasonal temperatures, which aids in the planning process for trade shows.

➤ A Stunning Cityscape

While a stunning cityscape isn't a requirement, it definitely makes exhibits in Las Vegas attractive. The cityscape looks incredible during the day and at night, it's even more beautiful. It's always great to have a nice backdrop for your trade shows, and the gorgeous cityscape of Las Vegas is perfect.

➤ Incredible Dining Options Throughout the City

Trade show attendees need great dining options, and Las Vegas offers incredible dining options throughout the city. From casual fare to fine cuisine, the culinary landscape is packed with options for everyone. All-you-can-eat buffets abound, many at great prices, and yet you can also find high-end restaurants with some of the best chefs in the world in this city as well.

➤ Las Vegas Generates Strong Attendance

Las Vegas is known for its casinos, grand cityscape, attractions, and great hotels, and it's a combination of these things that generates strong attendance for events held in the city. If you want to make sure everyone shows up to your event, then exhibits in Las Vegas are the perfect option.

➤ Plenty of Trade Show Services

If you want to ensure that you have plenty of trade show and convention services, Las Vegas won't disappoint. The city has thousands of resort industry employees who are experts at hosting conventions, meetings, trade shows, and more. You'll get the services you need from employees that work in a city dedicated to offering great services to business and leisure travelers.

➤ Activities and Entertainment Day and Night

Of course, you can't ignore the fact that Las Vegas offers activities and entertainment day and night, another great reason to choose this city for your trade show. During down time after trade show exhibits in Las Vegas, everyone will find plenty to do and see.

It's easy to see why so many trade shows take place in Las Vegas each year. If you're heading to a Las Vegas trade show, make sure you're well prepared with the best trade show displays rental so you can stand out from other trade show exhibits in Las Vegas.

Why Las Vegas Is thetop Destination for trade shows and getting business done in 2022? Vegas is more than just the Entertainment Capital of the World. Offering state-of-the art venues, culinary options and accommodations—with safety as a priority—the city is moving forward as a destination where business is happening.

The incomparable energy and creativity of Las Vegas make it an inspirational destination, one that's ideal for business professionals to gather, network and collaborate. Not to mention the new venues and innovative meeting spaces, unparalleled tech capabilities, dedication to sustainability and team-building potential. But that's not the only reason it's been the No.1 trade show destination for decades.

While the pandemic slowed down the world, Vegas continues to be a sought-after destination, one where health, safety and hosting a successful meeting or trade show exhibition are a prime focus. What else does Vegas have to offer professionals? Here are the top four reasons why Vegas is the place to do business:

1. Business happens in person, and it's happening in Las Vegas

How do we know? Well, the Las Vegas Convention Center alone hosted 52 large-scale events with nearly 750,000 attendees in 2021. With studies showing that employees are looking to reconnect with their colleagues and rebuild the chemistry of collaboration, in-person events and meetings will continue to rebound and grow. According to the Las Vegas Business Traveler Quantitative Research Overview, 91% of business travelers say they miss in-person events, while Freeman Research reports that 85% of American workers say that in-person meetings are irreplaceable. For all those attendees returning or visiting for the first time, Las Vegas is a destination with abundant accommodations that meet a range of price points, as well as offering a safe and lively atmosphere to make those interactions a reality.

2. Las Vegas is moving forward

Very recently, Las Vegas added notable venues to its impressive portfolio of spaces to host memorable meetings and events. New builds include the highly anticipated Resorts World that opened in June 2021; Virgin Hotels Las Vegas, which opened last March; Circa Resort and Casino, Caesars FORUM and The Stella Studio at The Venetian, each of which opened in October 2020; and the noteworthy West Hall expansion at the Las Vegas Convention Center (LVCC) which was completed in early June.

With these new venues opening, plus expansions and development happening at existing venues and consistent implementation of state-of-the-art tech, Las Vegas continues to grow—without capacity limits. Among the many venues available, there are flexible event and meeting space offerings as well as many outdoor spaces designed to move business forward. From large trade show floors to smaller, more intimate spaces that meet all budgets, there are accommodations and venues for all attendee and event types.

3. Getting around has never been easier or more convenient

Travel is insanely convenient with flights coming in and out at the nearby Harry Reid International Airport. When traveling around the city, the Las Vegas Monorail provides direct access to seven properties including the convention center, plus there are plentiful ride-share, taxi and rental car options. And don't forget the innovative underground transportation system, the Las Vegas Convention Center Loop, which provides fast and convenient transportation across the LVCC campus. On and off the clock, there's so much to see and do, any employee will be happy to do business in Las Vegas.

4. Meet Smart, Vegas Smart

Though COVID-19 still looms, Vegas venues adhere to all health and wellness guidelines that come from the CDC and the state, and are prepared to go above and beyond to fulfill individual event requests for enhanced protocols. That's why Las Vegas has been able to host many in-person events in the last year while keeping safety a top priority. Event spaces, facilities, airports and airlines undergo rigorous screenings to ensure high standards of health and hygiene are in place. The Las Vegas Convention Center, Mandalay Bay Convention Center and The Venetian Convention and Expo Center are all GBAC STAR™ accredited facilities, adding to the commitment to provide safer events in Las Vegas.

(This article is edited according to Internet information)

 NEW WORDS AND PHRASES

anchor　v. 支持, 使稳固

accelerate　v. 加速, 增加

strain　n. 负担, 困难

humidity　n. 湿度

gallery　n. 走廊, 画廊

limousine　n. 豪华轿车

culinary　adj. 烹饪的, 食物的

sought-after　adj. 受欢迎的, 吃香的

insanely　adv. 疯狂地

loom　vi. 隐约出现

adhere to　坚持

protocol　n. 医疗方案

rigorous　adj. 严密的, 严格的

accredited　adj. 官方认可的, 鉴定合格的

 **Reading 3**

Dubai

Dubai is the most populated city in the United Arab Emirates (UAE) and the capital of the Emirate of Dubai, the most populated of the 7 monarchies which together form the United Arab Emirates. Established in the 18th century as a small fishing village, the city grew rapidly in the early 21st century with a focus on tourism and luxury, having the second most five-star hotels in the world, and the tallest building in the world, the Burj Khalifa, which is over a half a mile tall.

In the eastern Arabian Peninsula on the coast of the Persian Gulf, it is also a major global transport hub for passengers and cargo. Oil revenue helped accelerate the development of the city, which was already a major mercantile hub. A center for regional and international trade since the early 20th century, Dubai's economy relies on revenues from trade, tourism, aviation, real estate, and financial services. Oil production contributed less than 1 percent of the Emirate's GDP in 2018. The city has a population of around 3.4 million (as of 2021).

During the 1970s, Dubai continued to grow from revenues generated from oil and trade, even as the city saw an influx of immigrants fleeing the Lebanese civil war. Border disputes between the Emirates continued even after the formation of the UAE; it was only in 1979 that a formal compromise was reached that ended disagreements. The Jebel Ali port, a deep-water port that allowed larger ships to dock, was established in 1979. The port was not initially a success, so Sheikh Mohammed established the JAFZA (Jebel Ali Free Zone) around the port in 1985 to provide foreign companies unrestricted import of labor and export capital. Dubai airport and the aviation industry also continued to grow.

The Gulf War in early 1991 had a negative financial effect on the city, as depositors withdrew their money and traders withdrew their trade, but subsequently, the city recovered in a changing political climate and thrived. Later in the 1990s, many foreign trading communities—first from Kuwait, during the Gulf War, and later from Bahrain, during the Shia unrest—moved their businesses to Dubai. Dubai provided refueling bases to allied forces at the Jebel Ali Free Zone during the Gulf War, and again during the 2003 Invasion of Iraq. Large increases in oil prices after the Gulf War encouraged Dubai to continue to focus on free trade and tourism.

Dubai Creek runs northeast-southwest through the city. The eastern section of the city forms the locality of Deira and is flanked by the Emirate of Sharjah in the east and the town of Al Aweer in the south. The Dubai International Airport is located south of Deira, while the Palm Deira is located north of Deira in the Persian Gulf. Much of Dubai's real-estate boom is concentrated to the west of Dubai Creek, on the Jumeirah coastal belt. Port Rashid, Jebel Ali, Burj Al Arab, the Palm Jumeirah and theme-based free-zone clusters such as Business Bay are all located in this section. Dubai is notable for sculpted artificial island complexes including the Palm Islands and The World archipelago.

Dubai has a hot desert climate. Summers in Dubai are extremely hot, prolonged, windy, and humid, with an average high around 40 °C (104 °F) and overnight lows around 30 °C (86 °F) in the hottest month, August. Most days are sunny throughout the year. Winters are comparatively cool, though mild to warm, with an average high of 24 °C (75 °F) and overnight lows of 14 °C (57 °F) in January, the coolest month. Precipitation, however, has been increasing in the last few decades, with accumulated rain reaching 110.7 mm (4.36 in) per year.

Alcohol sale and consumption, though legal, is regulated. Adult non-Muslims are allowed to consume alcohol in licensed venues, typically within hotels, or at home with the possession of an alcohol license. Places other than hotels, clubs, and specially designated areas are typically not permitted to sell alcohol. As in other parts of the world, drinking and driving is illegal, with 21 being the legal drinking age in the Emirate of Dubai.

Article 7 of the UAE's Provisional Constitution declares Islam the official state religion of the UAE. The government subsidizes almost 95% of mosques and employs all Imams; approximately 5% of mosques are entirely private, and several large mosques have large private endowments. All mosques in Dubai are managed by the Islamic Affairs and Charitable Activities Department also known as "Awqaf" under the Government of Dubai and all Imams are appointed by the Government. The Constitution of the United Arab Emirates provides for freedom of religion. Expats held preaching religious hatred or promoting religious extremism are usually jailed and deported.

The UAE culture mainly revolves around traditional Arab culture. The influence of Arab and Islamic culture on its architecture, music, attire, cuisine, and lifestyle is very prominent as well. Five times every day, Muslims are called to prayer from the minarets of mosques that are scattered around the country. Since 2006, the weekend has been Friday and Saturday, as a compromise between Friday's holiness to Muslims and the Western weekend of Saturday and Sunday.

Tourism is an important part of the Dubai government's strategy to maintain the flow of foreign cash into the Emirate. Dubai's lure for tourists is based mainly on shopping, but also on its possession of other ancient and modern attractions. As of 2018, Dubai is the fourth most-visited city in the world based on the number of international visitors and the fastest growing, increasing by a 10.7% rate. The city hosted 14.9 million overnight visitors in 2016, and is expected to reach 20 million tourists by 2020.

A great tourist attraction in Dubai is the Burj Khalifa, currently the tallest building on Earth. Although, Jeddah Tower in Jeddah, Saudi Arabia is aiming to be taller. Dubai has been called the "shopping capital of the Middle East". Dubai alone has more than 70 shopping centers, including the world's second largest shopping center, Dubai Mall. Dubai is also known for the historical souk districts located on either side of its creek. Traditionally, dhows from East Asia, China, Sri Lanka, and India would discharge their cargo and the goods would be bargained over in the souks adjacent to the docks. Dubai Creek played a vital role in sustaining the life of the community in the city and was the resource which originally drove the economic boom in Dubai. As of September 2013, Dubai creek has been proposed as a UNESCO World Heritage Site. Many boutiques and jewelry stores are also found in the city. Dubai is also referred to as "the City of Gold" as the Gold Souk in Deira houses nearly 250 gold retail shops.

On 2 November 2011, four cities had their bids for Expo 2020 already lodged, with Dubai making a last-minute entry. The delegation from the Bureau International des Expositions, which visited Dubai in February 2013 to examine the Emirate's readiness for the largest exposition, was impressed by the infrastructure and the level of national support. In May 2013, Dubai Expo 2020 Master Plan was revealed. Dubai then won the right to host Expo 2020 on 27 November 2013.

The main site of Dubai Expo 2020 was planned to be a 438-hectare area (1,083 acres), part of the new Dubai Trade Centre Jebel Ali urban development, located midway between Dubai and Abu Dhabi. Moreover, the Expo 2020 also created various social enlistment projects and monetary boons to the city targeting the year 2020, such as initiating the world's largest solar power project.

The Dubai Expo 2020 was scheduled to take place from 20 October 2020 until 10 April 2021 for 173 days where there would be 192 country pavilions featuring narratives from every part of the globe, have different thematic districts that would promote learning the wildlife in the forest exhibit too many other experiences.

Due to the impact of COVID-19 the organizers of Expo 2020 postponed the Expo by one year to begin in 2021（the new dates are 1 October 2021–31 March 2022）.

The Burj Al Arab, a luxury hotel, is frequently described as "the world's only 7-star", though its management has never made that claim but has claimed to be a "five-star deluxe property". The term "7-star hotel" was coined by a British journalist to describe their initial experience of the hotel. A Jumeirah Group spokesperson is quoted as saying: "There's not a lot we can do to stop it. We're not encouraging the use of the term. We've never used it in our advertising." The hotel opened in December 1999.

Burj Khalifa, known as the Burj Dubai before its inauguration, is an 828-meters（2,717 ft） high skyscraper in Dubai, and the tallest building in the world. The tower was inspired by the structure of the desert flower Hymenocallis. It was constructed by more than 30 contracting companies around the world with workers of a hundred nationalities. It is an architectural icon, named after Sheikh Khalifa bin Zayed Al Nahayan. The building opened on 4 January 2010.

The Palm Jumeirah is an artificial archipelago, created using land reclamation by Nakheel, a company owned by the Dubai government, and designed and developed by Helman Hurley Charvat Peacock/Architects, Inc. It is one of three planned islands called the Palm Islands which extend into the Persian Gulf. The Palm Jumeirah is the smallest and the original of three Palm Islands, and it is located on the Jumeirah coastal area of Dubai. It was built between 2001 and 2006.

Dubai International Airport（IATA: DXB）, the hub for Emirates, serves the city of Dubai and other Emirates in the country. The airport is the third-busiest airport in the world by passenger traffic and the world's busiest airport by international passenger traffic. In addition to being an important passenger traffic hub, the airport is the sixth-busiest cargo airport in world, handling 2.37 million tons of cargo in 2014. Emirates is the national airline of Dubai. As of 2018, it operated internationally serving over 150 destinations in over 70 countries across six continents.

The Emirati attire is typical of several countries in the Arabian Peninsula. Women usually wear the "abaya", a long black robe with a hijab（the head-scarf which covers the neck and part of the head-all of the hair and ears）. Some women may add a niqab which cover the mouth and nose and only leaves the eyes exposed. Men wear the "kandurah" also referred to as "dishdasha" or even "thawb"（long white robe）and the headscarf（ghotrah）. The UAE traditional ghutrah is white and is held in place by an accessory called "egal", which resembles a black cord. The younger Emiratis prefer to wear red and white ghutrah and tie it around their head like a turban. The above dress code is never compulsory and many people wear western or other eastern clothing without any problems, but prohibitions on wearing "indecent clothing" or revealing too much skin are aspects of the UAE to which Dubai's visitors are expected to conform, and are encoded in Dubai's criminal law. The UAE has enforced decency regulations in most public places, aside from water parks, beaches, clubs, and bars.

Dubai is one of the most popular business and investment destinations in the Middle East. Trade Centers and Dubai Exhibition Center provide businesses with a lot of exposure to expand their audiences and opportunities. The primary goal is to provide the highest level of service to organizers, exhibitors, and visitors by offering a wide range of opportunities to all exhibitors.

➤ From beacon of possibility to global business hub: Dubai World Trade Centre at 40.

As Dubai World Trade Centre celebrates its 40th anniversary, Mahir Julfar, Senior Vice President—Venue Services Management, charts the venue's remarkable contribution to Dubai, to the region and to the world.

When Dubai World Trade Centre (DWTC) opened in 1979 it was a beacon of possibility, a physical manifestation of the Emirate's ambition, and the bold vision of its leadership. In the intervening four decades, DWTC has become the heart of the city as it has grown and become a platform for opportunity, innovation and connectivity.

This year, as we celebrate our 40th anniversary, DWTC enjoys an unparalleled reputation as the region's premier venue for events, exhibitions, congresses and conferences. And yet that only tells half the story: today DWTC helps to position Dubai as a leading global business hub; it is a gateway for regional market access; and it generates considerable multiplier impact for the Emirate's GDP. Our story has grown in so many ways over the last 40 years. And there are many chapters still to come.

A gateway to the region, and the world. Much of that business stretches beyond the borders of the Emirate, the country, and even the region. Just as we welcome organizers, exhibitors and visitors from all over the world, so the progress that is made in our halls and concourses has impact in markets both established and emerging.

Today, Dubai is a gateway into fast-growing economies in the Middle East, Asia and Africa, as well as a preferred destination for businesses from Europe and the West. We are an active and much-valued facilitator of intercontinental trade and development. And our role extends far beyond the encouragement of domestic export/import markets-we are a true regional and global business hub and a catalyst for growth, bringing the world together and connecting possibilities everywhere.

As a city, Dubai is a hugely attractive destination for Foreign Direct Investment, and DWTC has always given prospective foreign investors the opportunity to come and see for themselves what the Emirate can offer: excellent infrastructure, a large number of fast-growing industries, asupportive government and regulatory environment-and the ability to reach into high-potential markets with fast-rising populations and rapidly accelerating needs.

Thousands of jobs, billions of dollars. DWTC attracts all visitor types to Dubai, with over 3.4 million people attending more than 600 events per year. In 2018, we hosted 97 large-scale business events, attracting more than 1 million international attendees, amounting to 41 percent of total attendance. This is a huge number, one that not only has a significant impact on the Emirate's events community and a plethora of associated industries, but also contributes to the Emirate's overall GDP.

Our organizers, exhibitors and visitors all have a direct, indirect and induced impact on spending across the city. Contributing to the first pillar of our impact model, direct spend covers a range of sectors, including hotels and restaurants, retail, transport, and government and business services.

At the same time, those industries that supply raw materials, manufactured goods and ancillary services all see a surge in demand for their products. This supply chain-which features purchases such as utilities, food & commodities, and construction has an indirect impact that ripples across the city and its industries, driving growth and generating incremental value in the economy.

That ripple effect contributes to the third and final area of our impact model: induced impact. This is the effect on the economy of the increases in household income driven by activity here at DWTC. That's consumption-induced impact that manifests itself in higher spending on food, consumer products, retail & entertainment, transport, housing, financing and myriad other goods and services.

In 2018, the combined value of these direct, indirect and induced impacts from those 97 large-scale events amounted to AED 23.0bn ($6.3bn). Gross Value Added to the Dubai economy by DWTC was AED 13.1bn ($3.6bn). In addition, we contributed 3.3 percent to Dubai's GDP over the calendar year, and supported nearly 88,000 jobs in the Emirate-an increase of 4.3 percent on 2017.

Turning innovative thinking into tomorrow's reality. Just as our growth has echoed and underpinned that of Dubai, so our future will mirror the remarkable changes in how we live, work and play in tomorrow's society.

We don't have to look far into our collective future to see how technologies such as autonomous transport, fintech, artificial intelligence and renewable energy solutions might enable us to build an even bolder and more sustainable tomorrow. DWTC and its trade events has become known across the region as a magnet for creativity and innovation in these fields, welcoming thought leaders and change-makers, and continuing to cultivate, support and drive their innovation: from our halls, to your homes. Through our exhibitions, conferences and trade shows, we act as an incubator of ideas. Across multiple sectors and industries, we facilitate debate, we enable partnerships and we provide a platform and forum for everyone: individuals, nations, organizations and companies-to articulate, demonstrate and participate in shaping tomorrow.

And innovative thinking is not just the domain of our speakers, delegates and presenters. Over the coming months and years, we will continue to welcome top talent from around the world as well as local startups to our growing One Central real estate development, which offers leading-edge facilities at the very best location in Dubai.

Our status as a free zone means that companies that are launching or expanding into the heart of the Middle East's most cosmopolitan city can avail 100 percent foreign ownership and the tax-free repatriation of profits. They can also enjoy a well-regulated business ecosystem to compete even more effectively both regionally and globally. We constantly enhance and evolve our offering, so that tenants can focus on doing what they do best: generating sales, hiring the right talent and, ultimately, growing their business as quickly and as efficiently as possible.

Forging 40 more years of excellence. Excellence does not come as standard in any industry, particularly the events sphere, which is changing constantly according to the demands of organizers, exhibitors and visitors from around the globe. To have built and sustained a reputation for quality across four decades therefore requires talent and tenacity.

Here at DWTC, we are committed to delivering the very best experience for everyone who steps over our threshold. While we are all proud of what we have achieved, we know that the best way to equip ourselves for the future is to listen to those organizers, exhibitors and visitors whose job it is to imagine and deliver that future. Inspired and empowered by the vision of the leadership of Dubai, we are determined to help lead the development of the technologies and industries of tomorrow. Come and see for yourself: the future is here.

(This article is edited according to Internet information)

 NEW WORDS AND PHRASES

monarchy   n. 君主国

mercantile   adj. 商业的, 商人的

aviation   n. 航空, 飞机制造业

influx   n. (人或物的) 大量涌入

Emirates   n. 阿联酋航空公司

refuel   v. 补充燃料

cluster   n. (聚集在同一地方的) 一群人

sculpted   adj. 雕塑般的

humid   adj. 潮湿的, 湿热的

precipitation   n. 冰雹

mosque   n. 清真寺

endowment   n. 捐款, 捐赠

attire   n. 服装

minaret   n. 尖塔

creek   n. 小溪, 小湾

boutique   n. 时装店

hectare   n. 公顷

enlistment   n. 招募, 征募

pavilion   n. 临时建筑

inauguration   n. 开始, 开创

archipelago   n. 群岛, 列岛

reclamation   n. 开垦

compulsory   adj. 被强制的

indecent clothing   不雅的衣服

manifestation   n. 示威运动

gateway   n. 通道, 门户

ancillary   adj. 从属的

ripple   v. 如波浪般起伏

incremental   adj. 增加的, 增值的

repatriation   n. 遣送回国

ENGLISH FOR WORKPLACE COMMUNICATION

Sample Dialogue: Booth reception.

Situation: The following conversation is between Chen (A), the sales manager from Rainbow company, and Mr. Smith (B), a potential client.

A: Good morning. Welcome to our booth.

B: Good morning.

A: I'm Chen, from Rainbow company, and this is my business card.

B: Hmm, Rainbow company. Glad to see you. Could you tell me something about your company?

A: My pleasure. Our company was established in 2009. It has been engaged in design, research and development, manufacturing and marketing of TVS.

B: It sounds good.

A: Which product are you interested in?

B: High definition television.

A: Oh, I'd like to show you our new product—3D TV.

B: It looks so big. How much is it?

A: 7,000 dollars. Of course, I can give you some discounts.

B: Great! Would you please tell me more about it?

A: Yes! This television has a high sales volume in China, the most important is that it is a 3D TV, if you wear glasses can see dynamic image.

B: I'm really interested in it. I want to have one.

A: Thank you very much. This is a brochure about our company and products.

B: Thank you.

A: By the way, if you can place an order before the end of May, we will give you a special offer.

B: OK, I got that. I will consider your products after discussing with our team.

A: Thank you. Looking forward to cooperating with you.

B: Me too.

A: Have a good day!

B: You too. Goodbye.

参考答案

UNIT 1

📖 展会知识

1. 展览会的名称，主办方，展览面积，展馆，参展商，展品及观众来源等。

（1）CBME 为新英富曼集团（Informa Markets）旗下展会（包括 CBME 孕婴童展及 CBME 童装展，CBME 孕婴童展为全球领先孕婴童展，CBME 童装展为国内领先童装展）。2021 年上海展，国家会展中心，7 月 14—16 日，展会面积 30 万平方米，2 726 家优质展商，4 253 个品牌，97 504 位独立观众人数。

（2）塔苏斯博览集团（Tarsus Group）是一家集展览、会议、教育、杂志出版和在线媒体于一体的国际 B2B 传媒集团，总部位于伦敦，业务遍及全球多个垂直领域，包括航空、医疗、标签、旅游和制造等。

（3）"2021 年墨西哥蒙特雷工业机械制造展览会（EXPO MANUFACTURA）"于 2021 年 2 月 2—4 日在墨西哥蒙特雷 AC 国际商务中心举行，该展由美国克劳斯公司（EJK）举办，今年举办第 23 届，展览会一年一届，根据主办单位数据统计，共有来自中国、德国、法国、意大利、英国、美国、加拿大、日本等 44 个国家和地区的 351 家企业参展，展览面积超过 17 000 平方米。吸引专业观众来自 88 个国家和地区的 13 000 人次。

（4）2022 年美国奥兰多沙滩及水上运动用品展览会（Surf Expo），展会地点：美国奥兰多佛罗里达奥兰治县会议中心，主办方：Emerald Exposition。展会展出了 4 000 多个品牌，吸引来自美国、加勒比海地区、中美洲、南美洲及其他地区的 28 600 名专业买家、采购商及零售商，超过 2 500 个参展商完整地展示了包括服装秀及每年的颁奖典礼等活动。

（5）第 40 届棕榈滩国际船艇展（Palm Beach International Boat Show，PBIBS）于 2022 年 3 月 24 日到 27 日举办，展会有近 500 家参展商、超过 400 艘游艇和 22 场研讨会，为行业的各个领域提供丰富的体验。

（6）2021 年美国拉斯维加斯礼品及纪念品展 Souvenir and Resort Gift Show，主办方：Urban Expositions，展会总面积 30 000 平方米，参展企业 570 家，来自中国、韩国、日本、德国、俄罗斯、中国台湾、南非、澳大利亚、巴西、印度等，参展人数达 38 500 人。

（7）2020 年澳大利亚悉尼礼品展览会（Reed Gift Fairs Sydney），展览日期为 2020 年 4 月 17—22 日，展览地点为悉尼国际会展中心，主办单位为励展集团，观众人数达 8 400 人。

2. 国际展览会的作用有：展览会是促进买卖双方成交的场合，是发布新产品和新信息的舞台，是企业学习各国先进技术和产品设计的课堂，是同业之间交流的平台，是企业进行广告宣传的途径（答案不限于此表述）。

3. UFI 是国际展览联盟（UFI 是法语 Union des Foires Internationales 的缩写）的简称。在 2003 年 10 月 20 日开罗第 70 届会员大会上，该组织决定更名为全球展览业协会（The Global Association of the Exhibition Industry），仍简称 UFI。UFI 是迄今为止世界展览业最重要的国际性组织。UFI 通过密切会员之间的联系、发展会员主办的展览会等，进一步促进国际贸易，并通过研究会员遇到的问题，为会员提供交流信息和经验、探讨同行业发展趋势、加强合作、密切关系的机会。UFI 的主要任务是提高全球展览会举办水平，促进跨国界的产品交流，加强展览会服务业及展览会举办专业技能的交流。UFI 的核心任务是对国际性展会进行认证。

4. 略。

5. 所谓会展经济扩散效应又可以称为会展经济间接经济效益，是指通过举办会展活动对一方经济发展带来的环境改善、城市声誉提高、商品流动和生产要素重新组合配置方面的影响。会展行业在促进国际收支平衡、商品交易、技术和信息交流以及市场开拓等方面的作用已经得到广泛认可。经济发达国家和地区凭借场馆规模大、设施全、展会运作国际化程度高、知名品牌多等优势，通过大型国际会议和展览的举办，有力地促进了资本、技术、信息和商品的国际流动，对本国和本地区综合经济发展发挥着积极的推动作用。对悉尼的经济拉动作用可以体现在促进交通、酒店住宿、餐饮、购物、旅游、就业等。

6. 全球展览日：每年 6 月的第一个周三是"全球展览日"（Global Exhibitions Day，以下简称 GED），该活动由全球展览业协会（The Global Association of the Exhibition Industry，以下简称 UFI）和国际展览与项目协会（IAEE）等业内机构发起，旨在让各界人士更好地认识到展览业在推动经济发展方面的巨大作用。2022 年全球展览日是 6 月 1 日，2021 年是 6 月 2 日。

📖 Reading 1

Ⅰ. Answer the following questions.

1. A trade fair functions as part of the marketing mix, specifically, as part of the communications mix, as part of the price and conditions mix, as part of the distribution mix, and as part of the product mix.

2. The marketing mix consists of product design, adapting to price and conditions and the measures necessary for distribution and communication. These tools enable the company to exert an active influence on the sales market. When exhibitors take part in a trade fair, they can bring into play their company policies on communication, price and conditions, distribution and products.

3. Communication is one of the central functions of trade fairs and exhibitions. It is clear that the scope of participation in a trade fair, intensive contact between exhibitors and visitors can be achieved; Personal conversations between exhibitors and visitors have great value; a trade fair can convey much more vivid and active information about a product or service than any other component of the marketing mix; trade fairs are in terms of their value as a promotional spectacle and in terms of their availability to the exhibitor. Participation in a trade fair helps a company to reach more potential customers and to create a more favorable impression on existing customers. It is also possible to become aware of changes in the customer profile and in buying behavior more quickly

and more directly within the scope of participation in a trade fair.

4. Important aspects of the price and conditions mix include price, credit, discount, payment and service. Participation in a trade fair contributes towards a new conception of the existing price and conditions mix, and if desired, new areas of the market can be sounded out. The price and conditions mix must be arranged so that company aims can be achieved and company profits assured.

5. The distribution mix can be represented as follows: sales organization, distribution channels, storage and transport. Another consideration is whether the existing distribution channels need to be changed qualitatively or quantitatively.

6. Competitors are those who manufacture the same, or similar, products, including companies that use the same production processes, or offer substitutes for your company's products.

Ⅱ. Please translate the following English sentences into Chinese.

1. 市场营销可以理解为公司对现有和潜在市场所有活动的计划、协调和监控。

2. 贸易展览会也是大量销售线索的来源，这是任何公司销售策略的重要组成部分。

3. 参加贸易展会的另一个好处是有机会与老客户保持联系。

4. 然而，市场研究或公司企业设计中表达的视觉形象也可以包括在内。

5. 如果有意识地努力将展会与营销组合的其他要素协调起来，那么参加展会就能取得成功。

6. 参展在以下情况下通常是有意义的：公司的销售不限于一个地区；销售基于足够广泛的客户基础；产品或服务显示出高度的专业知识。

7. 这种竞争分析的目的可能是对你自己的市场地位进行更全面的评估。

8. 在贸易展会上所追求的目标始终来源于个人的营销目标。

9. 任何由此产生的订单被称为间接贸易展览会采购订单。

10. 这意味着参展商将根据其既定的目标选择一个合适的贸易展会，或根据现有的交易会改变其目标。

Ⅲ. There are 10 sentences in this section. Beneath each sentence there are four words or phrases marked A, B, C and D. Choose one word or phrase that best completes the sentence.

1. C

2. A

3. D

4. A

5. A

6. D

7. A

8. A

9. B

10. A

Ⅳ. Writing

（略）

UNIT 2

📖 展会知识

1. 汉诺威工业博览会（HANNOVER MESSE）始创于1947年，一年一届，是全球展出规模最大、技术含量最高的综合型工业技术贸易展，被公认为是联系全球工业设计、加工制造、技术应用和国际贸易的最重要的平台之一。发展至今，已经成为全球工业贸易领域的旗舰展、世界工业贸易的"晴雨表"和全球工业技术发展的风向标。

以"融合的工业——工业智能"为主题的2019年汉诺威工业博览会（工博会）于4月1日至5日举行。来自75个国家和地区的约6 500家参展商齐聚德国中部城市汉诺威，集中展示全球工业数字化转型等最新技术成果。

2. 可以。一从参展国来看，汉诺威工业博览会（HANNOVER MESSE）始创于1947年8月，经过半个多世纪的不断发展与完善，已成为当今规模最大的国际工业盛会，被认为是联系全世界技术领域和商业领域的重要国际活动。有越来越多的亚洲、美洲及非洲国家不远万里前来洽谈，使博览会成为一个真正的全球性的盛会，中国也是其中之一。二从展品范围来看，2021年工博会，其中动力传动技术展包括机械传动、轴承、齿轮、齿轮箱、制动器、链条、链轮、同步带、转向系统和转向轴、连轴节等。2022年工博会，其中集成自动化及动力、传动控制展包括滚动轴承、滑动轴承等各种轴承及附件。

3. 如果河北企业要参加汉诺威工博会，企业要了解展览会参展商的构成比例、具体的展出内容，企业开拓国际市场的目标，具体地，企业应该提前做好以下准备工作：

（1）制定与企业营销目标和整体商业战略相匹配的目标。

（2）告知客户企业参与情况。

（3）展台展示设计。

（4）尽早开始制作宣传材料。

（5）考虑海关和进口规定，以及样品产品所需的许可证。

（6）对展台员工进行培训。

4.（1）2017年德国汉诺威工业展共吸引了来自全球75个国家和地区的6 551家参展企业，20多万名专业观众，总面积达35万平方米，其中61%的参展企业来自德国之外的国家。（2）2018年德国汉诺威工业展与汉诺威物流展同期举办，两大展会共吸引了来自全球75个国家的5 800家企业参展，专业观众数量达21万人次，其中超过7万人次来自德国之外的国家，总展出净面积达23万平方米。本届展会共有来自中国的980家企业参展，中国观众人次达6 500人，占据海外专业观众首位。（3）2019年德国汉诺威工业博览会净展出面积22.7万平方米，有来自75个国家和地区的6 500多家厂商参展，参展观众为21.5万名，其中有超过8万名观众来自国外。仅次于德国观众，中国观众人数（7 200）位居第二，荷兰（5 900）第三，意大利（3 400）第四，美国（3 400）第五。合作伙伴国瑞典观众人数为2 600人。（4）2020年汉诺威工业博览会，展出面积为23万平方米，展商数量为6 550家（62个国家，52%为德国以外的国家），观展数量为22.5万名（25%为德国以外的国家）。

5. 中国国际工业博览会（以下简称工博会）是经国务院批准、国家级的以装备类机电产品为展示、交易主体的专业性、工业类展会。

日本名古屋工业展览会（Manufacturing World Nagoya）是日本最大规模、最具影响力的工业展之一，是集轴承、紧固件、机械弹簧、金属、塑料加工技术等各类机械零部件的展览会。

英国国际工业分包展览会（SUBCON）是国际分包领域的重要展览会之一，该展会是以承揽来样、来图加工及分包零部件加工为业务的专业展览会，每年举办一届，其展品范围涵盖整个加工制造业，包括各种材料的部件、电子产品及机械安装、设计及样机、特种加工、精加工、检测，以及为生产线提供零部件成品的加工服务等。

波兰工业分包展览会（EUROTOOL – EXHIBITION OF INDUSTRIAL SUBCONTRAC-TING），是用于材料加工的机床、工具、装置和设备的国际贸易展览会，展示工业展览会机床、工具金属加工行业分包、供应商和合作伙伴塑料、橡胶、复合材料。

美国芝加哥工业展览会（HANNOVER MESSE USA）是德国汉诺威工业分支展。是制造商发现世界领先技术的地方，这是北美最全面的工业技术展会。

澳大利亚国际机械制造展会（National Manufacturing Week，NMW）是大洋洲规模最大、水平最高、涉及范围最广的工业技术展示与交易场所，每两年一届，分别在悉尼和墨尔本两个城市轮回展出。该展为澳大利亚最大的机械工业展会，展览范围较广，基本囊括了机械工业方面的各种产品。

海湾（巴林）国际工业博览会是巴林最大的展览会，也是中东地区最全面和最重要的工业展览会之一，已经成功走入第4个年头，影响力辐射至沙特、阿联酋等重点海湾区域外，也吸引来自北非、欧洲、亚洲等国家的参展参观商。其领域覆盖铝加工、各类机床制造、工业设备、制造业、自动化、钢铁及金属加工、能源、港口、物流及自由贸易区（保税区）等。

📖 Reading 1

Ⅰ. Answer the following questions.

1. (1) You should identify your objectives and define your target audience.

(2) Budget should be taken into consideration.

(3) You should select the right marketing tools and leverage social media.

(4) You should implement your strategy and establish a timeline for its execution.

2. When companies choose the right shows, they should make sure the following points.

(1) They should make sure their products or services are ready and target their market and audience.

(2) The trade show selected should align best with their objectives and goals.

(3) They should know how this trade show complements your export strategy.

(4) They should make sure that they have the necessary resources.

(5) They should make sure whether they have the capacity to follow-up on new leads.

3. (1) Set goals that compliment your marketing objectives and overall business strategy.

(2) Inform clients of your participation.

(3) Designing a booth display.

(4) Begin developing promotional material early.

(5) Consider the customs and import regulations, as well as licenses needed for your sample

products.

（6）Conduct staff training.

4. Arrive in advance;

Set up booth;

Take time to understand the audience;

Spot genuine leads;

Follow the networking tips;

Manage the leads.

5. Professional follow-up within a recommended 30-day period helps to ensure your business reaps the most benefit from trade show participation. Initial follow-up can be made by a simple-phone call within a few days of the show, while more intensive follow-up should be made within the span of a month. Examples of follow-up include thank you letters or e-mails, trips to visit leads, and distribution of product samples and information packages.

Ⅱ. Please translate the following English sentences into Chinese.

1. 贸易会展让公司可以在一个地点与不同的潜在买家和客户群体互动并建立新的合作关系。

2. 确保用于分享和嵌入内容的社交媒体工具显示在显著的、具有战略意义的位置，并且无论是在宣传材料上还是在您的网站上，都易于访问。

3. 出口和商业目标越清晰，就越能确定哪个贸易展能给你的公司带来最大的价值，整个展览也就越有重点。

4. 它是一种营销工具，必须整合到你的整体商业策略中，为了获得想要的结果，必须正确地使用它。

5. 71%的公司表示，他们在国际上展出的主要原因是增加线索和销售，并与客户/潜在客户建立关系。

6. 为了帮助做好展会前的准备工作，并确保你在展会上充分利用你的时间，如果可能的话，提前预约潜在客户和买家是一个好主意。

7. 对于成功的贸易会展，创造新颖的、创新的与无与伦比的展示和销售技术是至关重要的。

8. 无论你是独立展出，还是作为一个大型展馆的一部分，重要的是要注意细节，以确保你所创造的印象和气氛有利于成功的商业活动。

9. 展台工作人员不仅应该能够回答与产品或服务有关的基本问题，他们还应该能够回答与公司的能力、出口意图、当前的市场风险和拓展其他市场的努力等相关的广泛问题。

10. 如果他们不能向与会者传递产品和公司的重要信息，也不能为潜在客户提供专业和胜任的代表，那么他们的努力就会白费。

Ⅲ. There are 10 sentences in this section. Beneath each sentence there are four words or phrases marked A, B, C and D. Choose one word or phrase that best completes the sentence.

1. B

2. D

3. B

4. C

5. A

6. C

7. D

8. A

9. B

10. C

Ⅳ. Writing

（略）

UNIT 3

展会知识

1. 德国法兰克福春季消费品展览会（Ambiente）由德国法兰克福展览公司主办，是全球最大的消费品展览会，于每年 2 月举行。该展览会是消费品行业内最具影响力的国际性展览会之一，是企业成功进入国际市场的重要贸易渠道。该展会不仅是各国参展商产品信息交流的中心，同时也是广大参展商结识新客户、同行之间交流、企业学习各国先进技术和宣传自身的理想场所。

展品范围：

（1）餐桌、厨具和家用品世界：玻璃、瓷器、陶瓷、金属器皿、烹饪器具、餐盘和油炸锅、小型电器、烘焙器皿、家用清洁设备及器皿、厨具趋势、厨具附件和纺织品、美食商店等。

（2）礼品世界：国际礼品系列、文具、皮革制品、游戏、工艺品及手工艺品、烟具、画、蜡烛、制造的和收集的限定版的礼品、手工雕刻系列的产品、原创设计和流行的礼品、季节性的装饰品、当代工艺品、服装和珠宝等。

（3）室内装饰世界：乡村住宅、民族家具和附件、有设计的家具、家居附件和纺织品、室内灯具、个人的家具、经典家居附件、画及画框、浇铸的镜子、家具、家用附件、家用纺织品、室内灯具等。

2. 铸铁壶、铸铁锅、不锈钢餐具、拖把、便携式缝纫机和针线盒、瓷器等。因为法兰克福消费品展有 170 多个国家和地区的采购商参与，是消费品行业最重要的展会，所以文中这些公司一定要参加这个展览会。

3. OEM 优势：

（1）规模生产，降低成本。

根据规模经济效应，外贸企业参与 OEM 可有效地扩大自身的发展规模，实现生产的集约化和规模化，不但增加了产品销路，也起到了降低成本的作用。

（2）学习经验，完善管理。

品牌生产者在产品开发、生产管理、市场营销等方面也具备较强的实力，OEM 模式为我国外贸企业在这些方面学习和借鉴国外经验提供了机遇，我国外贸企业在生产中学习，迅速提高自身的技术和管理水平。

（3）产品创新，提升能力。

我国外贸企业通过给不同品牌生产者代工从而提高自身的技术实力，增加自身生产线的长度和深度，更可以推出自己的品牌，参与到市场竞争之中。

OEM 劣势：

（1）OEM 生产模式容易使一部分外贸企业丧失核心竞争力。

OEM 品牌生产者控制着核心技术和销售网络，我国外贸企业如果一味地代工、沉溺于眼前的蝇头小利，不追求技术创新、完善管理和开拓自己的销售网络，随着行业竞争的加剧和中国劳动力成本的上升，最终会丧失企业的核心竞争力。

（2）OEM 生产模式不具备可持续性。

由于我国很多行业的生产环保标准都比欧美发达国家低，一定程度上吸引了 OEM 品牌生产者将高污染项目转移到中国，这使得中国的生态环境和人居环境加速恶化。随着国家和人们环保意识的觉醒，OEM 模式将难以持续下去。我国外贸企业如果过分依赖 OEM 模式，必将走入困境。

4. 一般来说，国际会展规模都很大，一些世界著名的国际展览会可以由几千家参展商参展，专业观众达到几十万人，国际会展一般持续 2~3 天，在同一时间、同一地点使某个行业或某个地区的重要生产厂家和购买者集中到一起，这在其他场合是办不到的。

因此，参加国际展会，企业可以在短时间内签单、获得有成交意向的客户或者获得大量对本公司产品感兴趣的潜在客户，回国后，有成交意向的客户稍加发展即可成交，对公司产品感兴趣的潜在客户，只要展后跟踪策略得当，也会有一部分转化为客户。因此参加国际展会是外贸企业促进销售的重要机会。

在一场展览会上影响参展企业结识客户，获取订单的因素如下：

（1）参展商在参展前的准备工作是否充分。

（2）参展商展位的设计是否能够吸引参观者。

（3）参展商的员工是否经过培训，具备良好的企业形象和沟通能力。

（4）参展商参展是否有连续性。

（5）参展商员工是否能够获得包含充分信息的销售线索。

📖 Reading 1

Ⅰ. Answer the following questions.

1. At least 40 percentage.

2. 10 ways.

3. The duty of Lead Sheriff is to observe staffer/attendee interactions in the exhibit and make sure the leads got recorded, along with any promises made.

4. The follow-up step is simply making sure the attendee knows that you have registered his or her request. It's the "Thank you for visiting our exhibit" email, and it should be done before the lead is distributed to sales.

5. 1st approach: inviting prospects to a webinar relevant to the show.

2nd approach: sending a direct mailer after the show but before the follow-up call from sales.

Ⅱ. Please translate the following English sentences into Chinese.

1. 如果你没有采用正确的方式跟踪客户，就是浪费钱。

2. 需要明确的是，所有公司都想跟踪潜在客户，但是缺乏系统的计划，所以经常会阻碍这些想法的实现。

3. 此外，多源信息建议给每个潜在客户评分，只把得分最高的客户交给销售部门。

4. 为了帮助销售代表消除打可怕的陌生电话的恐惧，要给他们充分的理由拿起电话，这些年，我已经实施了多种展会后跟踪潜在客户的方案。

5. 营销应当在展后评估潜在客户，把合格的客户交给销售，开始培育和发展不合格的客户。

6. 促进销售是公司划拨大笔资金参加展会活动的原因。

7. 如果你的营销团队没有跟踪潜在客户的时间表或者责任心，当潜在客户被忽视时，你不应该感到吃惊。

8. 如果从一个潜在客户身上不能挖掘到相关信息，包括他/她的需求、购买力、预算和时间等，这个客户信息就只能是一张商务名片。

9. 这必须处理并纠正，否则公司将因此继续遭受收益损失。

10. 和销售部门一起设计出一个合适的潜在客户评分模型，这个模型就成为你决定哪些客户应该舍弃、哪些应该培育、哪些应该立刻发给销售代表迅速跟踪的关键。

Ⅲ. There are 10 sentences in this section. Beneath each sentence there are four words or phrases marked A, B, C and D. Choose one word or phrase that best completes the sentence.

1. B
2. D
3. A
4. D
5. A
6. A
7. C
8. B
9. C
10. A

Ⅳ. Writing

(略)

UNIT 4

展会知识

1. 法国巴黎建筑展 BATIMAT 由世界最大的展览机构励展博览集团（Reed Exhibitions）主办，每两年一届。自 1959 年创办以来，规模日益扩大，影响力波及全球，已经发展成为法国境内最高端的行业展览会。2022 年 10 月举办的展会展览面积达到 30 万平方米，观众数量超过 30 万人，参展商数量超过 2 000 家。展品涉及建筑材料、门窗幕墙、装饰装修、智能化楼宇、建筑设备和工具、施工车辆及设备等方面。

2. 由于客户使用法语发邮件进行沟通，不用英语，为了能够维系客户以及和客户更好地沟通，李雪自学法语。经过一年的邮件沟通，李雪获得了客户的信任，在 2014 年的广交会上，两人再次见面，客户向李雪敞开心扉，提出了他心中的种种疑虑，李雪认真做了解释并带客户考察中国工厂，帮助客户全面了解护栏这个产品在中国的现状。李雪的努力使其获得了对方厚厚一本法语产品报价单。2015年，李雪与客户在法国建材展上再次碰面，经过两年的沟通和磨合，对方充分信任李雪，这使得李雪获得大订单。

3. 外贸企业连续参展很重要。

很多企业今天良好的展览效果不是通过一次参展行为偶然得到的，而是需要持续的参展，并选择更有潜力的展览项目，从而实现积累，筛选客户资源，扩大贸易份额。

另外，客户之间的信任和磨合不是一次参展能够建立起来的，经过连续参展，客户之间逐渐由生变熟，由试探了解变为信任，由初步合作成为几十年的老客户，由小客户发展为大客户，所以企业持续参展对于企业获得客户、培养和稳定客户很重要。

4. 展后追踪对于企业实现参展价值至关重要。

人们的记忆是个几何倍数衰减的过程，根据实践经验，展会结束3天内，大多数人都能记得展会期间交流过的人及交谈的内容，不仅是客户，还包括现场沟通的参展人员；展会结束后1周内，记忆内容减半；展会结束后2周，记忆内容减少高达七成，展会结束后1个月，只有在展会中非常特殊和重要的事情才能被记忆唤起。因此，在展会结束后及时进行展后跟踪，趁热打铁，在客户还没有忘记我们的时候创造再次沟通的机会，最大限度地将展会上获得的销售线索转化为企业的订单，转变为企业的效益，对于实现企业参展价值至关重要。

此外，展后跟踪要取得明显成效，系统性的方法手段、明细跟踪任务和奖惩措施也非常重要。

📖 Reading 1

Ⅰ. Answer the following questions.

1. Create a timeline for yourself to avoid missing any important details of the planning process. Start your plan months ahead.

Ensure you have clear goals and understand why you're participating in that particular trade show.

Get the word out.

2. post-event email.

3. Make sure you don't use the same email for different trade shows. Consider putting the name of the show in the subject line, and even personalize your email campaign.

4. Trade leads and contact information.

Coordinate your post-trade show email campaign.

Offer a giveaway together if you can make contact with this other business before the event-day.

5. Those who don't fit your target customer profile or wouldn't benefit from your product or service can still be great resources for your organization throughout future partnerships or just spreading the word about your brand.

When you and your team find an opportunity, these connections might be suited for, they're still familiar with the organization, which can save you time when you try to move them to the nest steps in your marketing funnel.

Ⅱ. Please translate the following English sentences into Chinese.

1. 我们也建议你制定一个时间表，以便按计划行事，不错过计划过程中的任何重要细节。

2. 在实施展后策略、跟进客户之前，你需要从展会摊位参观者那里收集一些信息。

3. 在你的摊位入口处放置宣传册架子毫无意义。

4. 如果你的最终目标是一个邮件跟踪客户，显然，在会展期间你的目标就是收集邮件地址和其他联系信息。

5. 甚至展后一两天你就会遥遥领先于竞争对手，他们可能花费几个星期才发出第一批展后沟通邮件。

6. 活动后发送邮件是初次接触并将潜在客户转化为客户的好办法。

7. 考虑把展会名称纳入邮件标题，甚至使你的邮件活动个性化。

8. 确保对使用过的市场营销资料、活动前和活动后发送的信函以及效果有记录。

9. 记录与您的贸易展览有关的所有事情，您可以对令人失望的活动进行调整，或者在结果非常糟糕的情况下彻底修正整个策略。

10. 提前规划，在展会前制定你的跟踪策略将有助于使展后跟进客户成为轻而易举的事。

Ⅲ. There are 10 sentences in this section. Beneath each sentence there are four words or phrases marked A, B, C and D. Choose one word or phrase that best completes the sentence.

1. C

2. A

3. C

4. C

5. C

6. C

7. C

8. A

9. A

10. A

Ⅳ. Writing

(略)

UNIT 5

📖 展会知识

1. 德国最重要的展会城市为柏林、杜塞尔多夫、法兰克福、汉堡、汉诺威、科隆、莱比锡、慕尼黑、纽伦堡和斯图加特等。

2. 德国著名的会展公司有汉诺威展览公司、杜塞尔多夫展览公司、科隆展览有限公司、慕尼黑展览集团、柏林展览公司、埃森展览公司、莱比锡展览公司、法兰克福国际展览公司、斯图加特展览公司等。

3. 为缓解会展行业遭受的损失，2021 年 1 月，总额为 6.42 亿欧元的"德国联邦拯救伞计划"获批，德国会展行业相关企业可申请补贴，对象包括会展设施所有者、经营者及中介机构，最高补贴额为其全部利润损失。此外，德国经济事务与气候行动部联合各联邦州，推出了总额为 6 亿欧元的保险项目，以帮助展会组织者应对疫情带来的次生灾害。与此同时，德国会展行业积极探索新的办展方式。德国展览业协会表示，展会组织者加速数字化转型，以维持与客户的联系，并提供新产品信息。《会议和活动晴雨表2021—2022》报告显示，2021 年，混合展会数量增长了 280%，线上虚拟展会活动增加120%。在参与人数方面，2021 年，参加混合展会活动的人数为 1 840 万，而在 2020 年只有 180 万人。

4. 讨论内容可以围绕线下展会与线上展会各自的优缺点对比进行，如线下展会的优点包括面对面的真实交流，更好地了解商家与产品，容易从展会人群中发掘潜在的客户，更容易开发大客户与长期客户等；线下展会的缺点是花费大、造价高，包括展位费、交通住宿、会展人员等支出，易受环境、天气、政策等诸多因素影响。相比而言，线上展会各项费用较低，展会可容纳的参与群体更广，不易受到周围环境影响，展会时间更灵活，展会的产品随时可供展览，展会信息数据生成较快，对周围环境影响小，展会信息可以长久保存，传播渠道丰富，绿色环保；线上展会的缺点主要是商家与客户间相互了解受限，不容易达成大项目或长期项目的买卖，线上浏览容易，但不易专注于具体商品，进而导致实际产出效益低，对于相关经济的拉动作用降低。

5. 分组问卷调查设计鼓励多样性与创新性，问卷设计可以从展会产品、主题、地点、人员、布置、交通预算、住宿预算、餐饮预算、客户分析、后期跟进、产品体验、产品推广、周围环境、人文背景等各类与会展相关的内容设计。

📖 Reading 1

Ⅰ. Answer the following questions.

1. Trade fairs have always been centers of knowledge – of information that is prepared, cultivated and placed in helpful contexts. As our society becomes ever more knowledge-based, information has become a crucial resource. Producing, selecting, filtering and channeling it has thereby become one of the most important activities of national economies. As a result, ever more conferences are being held in conjunction with exhibitions, and vice versa, as vibrant and immediate ways of conveying knowledge.

2. Advantages for Germany as a trade fair location include state-of-the-art facilities, market-oriented trade fair strategies, high international presence, high professionalism, leading service standards, excellent cost-benefit ratios and attractive regional trade fairs.

3. Approximately 100 exhibition organizers are active in Germany, around 40 of which handle international fairs. Around 58, 000 German companies are active exhibitors in the B2B segment. The percentage of decision-makers among all trade fair visitors is exceptionally high at 63%. Managing directors, board members and self-employed people from Germany make up 35% of trade visitors, and 73% of those from abroad. Close cooperation with a large number of service companies is what makes a trade fair successful.

4. Nearly two-thirds of the leading global trade fairs in different branches of industry are held in Germany. The country is the world's number one location for international trade shows. Despite the worsening economic conditions, they again grew slightly in Germany compared to their previous

events, with only the number of visitors remaining constant. The competitive position of trade fairs vis-à-vis other marketing media remains stable.

5. Close cooperation with a large number of service companies is what makes a trade fair successful. Of special note here are stand constructors, designers, event specialists and consulting companies, as well as shipping companies, stand personnel trainers, caterers and hotels. The stand construction companies in the FAMAB professional association alone post overall annual sales of around € 2 billion. AUMA and FAMA members organize more than 300 trade fairs a year in major growth regions outside Germany, especially in Asia, North America, South America and Eastern Europe and this number is growing.

Ⅱ. Please translate the following English sentences into Chinese.

1. 在此过程中，展览组织者更加全方位发展成为市场合作伙伴。

2. 贸易展包罗万象，覆盖了发达国家各领域的专业展览及发展中国家综合展。

3. 德国贸易展占到世界贸易市场10%的份额。

4. 在所有的市场手段中，贸易展有着最为广阔的功能形式。

5. 作为新产品的测试市场，（展会）也是市场调研的有效工具。

6. 随着我们的社会越来越以知识为基础，信息成为至关重要的资源。

7. 它们成为反映世界市场的交流与创新论坛。

8. 德国贸易展览的一大竞争优势是它的国际吸引力：展会将世界市场吸引到了德国。

9. 这些公司中的一些承担了每年世界上二十多场顶级贸易展会。

10. 此外，他们为特定的区域量身定做开发了新的贸易展会主题。

Ⅲ. There are 10 sentences in this section. Beneath each sentence there are four words or phrases marked A, B, C and D. Choose one word or phrase that best completes the sentence.

1. B

2. C

3. D

4. A

5. D

6. C

7. B

8. C

9. A

10. C

Ⅳ. Writing

This is an open-type writing assignment with no set layout. Any innovation in writing is encouraged. You may present your report to include the following points:

1. Executive summary about the frequent turnover in your working and the subsequent damages it causes.

2. Report introduction about epidemic situation, contractors involved, related responsibilities, financial compensation and possible solutions.

3. Detailed findings on specific time periods, accurate data about turnover frequency, wor-

king staff conditions and so on.

4. Recommendations on virtual expos, possible opportunities or solutions in the future.

5. Conclusion of the report.

6. Your name, position and time (at the end of the report).

Your response will be judged on the basis of the quality of your writing. Typically, an effective response will be above 200 words.

UNIT 6

📖 展会知识

1. 德国法兰克福国际汽配展览会（Automechanika）由德国法兰克福展览有限公司（Messe Frankfurt GmbH）主办。该展会创建于 1971 年，每两年一届，是全球最具规模的国际汽车零部件、工艺装备及相关工业展览会。这个展览会每届都吸引着数千家的国际企业参加。展品范围主要包括汽车维修、加油站设备、汽车零部件与驱动和传动系统、汽车电气和电子电气系统以及环保废弃物处理等。如今，Automechanika 已成为一个汽车行业卓越的、创新的企业交流的平台。

2. 论坛主题包括替代驱动系统、培训和专业发展、弹性供应链和电子商务。

3. 本题目为开放型题目，会展中产品创新角度可以从产品外观、产品功能、产品形态、产品体验、销售渠道、售后保障、产品宣传等各个方面探讨。

4. 举办国家和城市包括德国法兰克福、土耳其伊斯坦布尔、越南胡志明、阿联酋迪拜、阿根廷布宜诺斯艾利斯、中国上海、马来西亚吉隆坡等。

📖 Reading 1

Ⅰ. Answer the following questions.

1. The International Home + Housewares Show staged since 1906 is organized by the IHA. The modern housewares exposition was born in 1927. The Show moved to Chicago's Navy Pier in 1949 to accommodate the growing number of companies. In January 1971, the 54th International Housewares Show was back at the exposition center. In 1991, NHMA moved to new quarters in Rosemont. In 1997, the Show opened in the grand new South building of the McCormick Place complex. In 2004, IHA moved the trade show from its long-standing January date to a March timeframe and renamed it the International Home + Housewares Show in recognition of show's evolution to a home goods marketplace.

2. IHA is committed to maximizing the success of the home + housewares products industry. IHA provides the world-class marketplace, The Inspired Home Show, as well as facilitation of global commerce, executive-level member share groups and conferences, a wide range of international business development tools, housewares industry market data and information services, facilitation of industry standards, and more!

3. The Show was segmented in five expos, including Clean + Contain Expo, Dine + Décor Expo, Discover + Design Expo, Wired + Well Expo, The International Sourcing Expo.

4. There are two critical International Housewares Association educational summits: CHESS

(Chief Housewares Executive SuperSession) and the International Business Council's Global Forum.

5. Top housewares brands that were unable to participate in The Inspired Home Show 2022 are committing to returning to the Show in 2023. More than 100 exhibitors that sat out the 2022 Show have already submitted space applications for the 2023 Show, which is set for March 4–7 at Chicago's McCormick Place Complex. IHA will continue to work to make the 2023 Show as meaningful and productive as possible for the benefit of all attendees.

Ⅱ. Please translate the following English sentences into Chinese.

1. 现有展商一直在努力拓展展位空间，潜在的展商迫切地扣响参展之门。

2. 家庭用品展的首要意义是增加了人与人之间的接触和交流。

3. 1956 年 Navy Pier 只能容纳不足 649 家参展商。

4. 团结凝聚之时，我们的行业才最成功，能够有如此多的公司与品牌支持是一件很棒的事。

5. 1979 年，西馆开启了一项服务，为新的展商额外提供一天的展期服务来吸引购买者。

6. 每年零售商来到芝加哥发掘创新性产品，与当前供应商见面，寻求未来的合作机遇。

7. 整个销售理念是围绕着创新、特色与爱家。

8. 无论是休闲的夜晚还是丰盛的晚宴，消费者都会通过家庭娱乐来培养他们的个人品牌。

9. 国际家庭用品协会将继续努力使 2023 年的展会尽可能有意义和富有成效，以造福所有与会者。

10. 技术继续让消费者的生活更轻松——更轻松带来更多需求。

Ⅲ. There are 10 sentences in this section. Beneath each sentence there are four words or phrases marked A, B, C and D. Choose one word or phrase that best completes the sentence.

1. B
2. A
3. C
4. D
5. C
6. D
7. A
8. D
9. C
10. A

Ⅳ. Writing

This is an open-type writing assignment with no set layout. Any innovation in writing is encouraged. You may present your application letter to include the following points:

1. Salutation: when you don't know the name of the recipient: Dear Sir/ Madam (RrE) Ladies and Gentlemen (AmE); when you know the name of the recipient: Dear Mr/ Mrs/ Ms/ Miss Family Name.

2. Letter contents: briefly express your application purpose and inquire further information.

3. Endings: Yours faithfully(RrE)，Sincerely yours (AmE)；Sign the letter, then print your name and position under your signature.

Your response will be judged on the basis of the quality of your writing. Typically, an effective response will be above 200 words.

UNIT 7

展会知识

1. 上述案例集中反映了国际展览会中的展品侵权问题。国际展览会中常常涉及的知识产权纠纷大多出现在专利与商标两方面。专利与商标经过法定程序获得后，即受到法律保护，权利人享有排他权利。未经权利人许可，任何人不得擅自生产、使用、销售或进出口专利产品以及具有相同商标的产品，否则就构成侵权行为，会受到法律的惩处。

2. 完善企业知识产权服务体系；加强企业知识产权文化建设；增强企业自主创新能力；积极应对国际知识产权侵权之诉。

Reading 1

Ⅰ. Answer the following questions.

1. UFI is one of the most important international organizations, which provides a platform for members to exchange information and experience, discuss the development trend of the industry, strengthen cooperation and close relations.

2. Intellectual Property (IP) can be divided into two categories: (1) Industrial Property, which includes trademarks, patents, utility models and designs; (2) Copyright and neighboring rights, which includes literary and artistic works such as novels, poems and plays, films, musical works, artistic works such as drawings, paintings, photographs and sculptures, and architectural designs.

3. A trademark identifies and distinguishes the products or services of a company from those of its competitors.

4. Patents cover inventions and protect the (technical) characteristics of a product or a process.

5. The requirements for the granting of a patent are in most countries, there must be a non-obvious technical contribution to the state-of-the-art. The state-of-the-art is formed by everything already known to the public before the filing of a patent application (even disclosures by the patent applicant themselves) .

6. The owner of an IP right can enforce that right through different actions – warning/request to discontinue use letters to infringers; customs actions; court actions; dispute-settling by mediation and arbitration. The collection of proof (in that case and in many countries, a "descriptive seizure" procedure or ("AntonPiller" order)) enables the IP right owner to obtain the decision through the Courts to send a neutral expert to the premises (offices, exhibitions) of an alleged infringer in order to describe the alleged infringement and seize evidence.

Ⅱ. Please translate the following English sentences into Chinese.

1. 展会提供了获取竞争对手信息与发现新产品和服务的绝佳机会，从而可在大规模制造和商业化之前的初始阶段查明潜在的知识产权侵权行为。

2. UFI，即全球展览业协会。它的使命之一是帮助其成员和展览业捍卫商业利益，同时促进展览成为最强大的营销、销售和沟通工具。

3. 商标的保护期通常为 10 年，可以续期。

4. 商标的保护在地域层面上有效（即，专有权仅存在于商标注册国）。

5. 整个欧盟（目前有 27 个成员国）都可以享受单一保护。

6. 专利期限通常为 20 年，并需缴纳费用（通常为每年一次）。还可以更新专利。

7. 专利提供的保护在地区层面上有效。在欧洲，通过一种语言的一次考试，可以使用一种程序（目前最多可为 37 个国家提供保护）。

8. 参展商在展会开始前保护和注册商标、专利或设计，以获得有效权利（展会破坏新颖性），因此在一般情况下和展会期间都使用各种形式的法律保护。

9. 应鼓励参展商声明其产品或服务受知识产权保护（如适用）。

10. 主办方应能够提供中立的仲裁员或法官，以帮助确定展会期间是否存在侵权行为或解决知识产权纠纷，并应提供口译员，以便在与外国参展商发生纠纷时进行沟通。

Ⅲ. There are 10 sentences in this section. Beneath each sentence there are four words or phrases marked A, B, C and D. Choose one word or phrase that best completes the sentence.

1. D

2. A

3. D

4. B

5. A

6. C

7. A

8. C

9. B

10. C

Ⅳ. Writing

（略）

UNIT 8

📖 展会知识

1. BAUMA 展会，俗称宝马展。全球工程机械行业规模最大的展示平台，第 33 届德国慕尼黑国际工程机械、建筑机械、矿山机械、工程车辆及零部件博览会（BAUMA2022）于 2022 年 10 月 24—30 日在德国慕尼黑市展览中心举行，展品范围将涵盖土石方机械、起重机械、筑养路机械、混凝土机械、掘进与凿岩机械、工程车辆、建材机械、矿山机械、脚手架及模板以及工程机械零配件等，展期共 7 天。有 280 家中国展商参展，展位净面积近 2 万平方米，是第二大国际展团。其中，由中国展团组织国内工程机械行业大、中、型骨干企业参展的"中国展团"面积达 15 316 平方米，占国内出展总面积的 77%。"中国展团"在 BAUMA 博览会上的出展规模逐届升级，扩大了中国企业在国际市场的影响力，提升了民族品牌产品的竞争力。

2. 卡特彼勒（Caterpillar）成立于 1925 年，总部位于美国伊利诺伊州。是世界上最大的工程机械和矿山设备生产厂家、燃气发动机和工业用燃气轮机生产厂家之一，也是世界上最大的柴油机厂家之一。

维特根集团（Wirtgen Group）始于 1961 年，成立于德国，全球知名工程机械品牌，主要生产和销售路面铣刨机、冷再生机、土壤稳定机、滑模摊铺机、露天采矿机等产品，致力于提供用于筑养路、目标矿物的开采和加工、建筑材料再生以及沥青生产的理想型解决方案。

利勃海尔（Liebherr）集团在全球范围内以工程机械设备、冷藏冷冻设备和铁路航天设备而著名。利勃海尔专注于制冷领域，产品包括奢华冰箱、葡萄酒柜、雪茄柜等产品。

3. 根据宝马展的展品范围，我国徐工集团、三一重工、中联重科、柳工集团、山河智能、中铁重工、浙江鼎力、北京盛瑞达国际展览有限公司等可以参加宝马展，展会预算略。

4. 俄罗斯工程机械宝马展览会（BAUMA CTT Russia）是俄罗斯以及中亚地区最大的专业行业展，更是中国工程机械企业开发俄罗斯、东欧地区最佳贸易平台。往届知名参展的国际代表企业有：利勃海尔、阿特拉斯、维特根、约翰迪以及临工、柳工、福田雷沃等。

美国拉斯维加斯工程机械展览会（CONEXPO—CON/AGG）在美国拉斯维加斯国际会展中心举办。该展览会是世界三大工程机械展之一，仅次于 BAUMA 的世界第二大工程机械展。展会规模宏大、客商众多，集中了世界知名品牌（如卡特彼勒、小松、利勃海尔、沃尔沃），是业内展示最新技术、设备和产品的重要平台。

法国巴黎工程机械展览会（INTERMAT）在法国北郊维勒班展览中心举办，该展览是全球第三大的工程机械展品交易及展示平台。在该展中，中国展团成为仅次于意大利、德国的第三大国际展团。

📖 Reading 1

Ⅰ. Answer the following questions.

1. Familiarize yourself with the state of the market. Know what vendors are charging and what items are commonly negotiable. Next, identify which issues are most important to you. Finally, figure out your acceptable bottom line ahead of time, and be ready with another plan.

2. Pay attention to show management; exhibit houses; transportation carriers; installation and dismantle contractorsand other vendors.

3. Such as audiovisual contractors, talent companies, promotional-products suppliers, and furniture-rental companies.

4. Ask about providing your own cabling/surge protectors, request that the electrician lay your electrical on straight-time hours, and find out if it will save you money to build a power distribution unit (electrical box) to break down large quantities of power to cheaper, smaller outlets.

5. If you are using an EAC, you can ask it to reduce its city rate to match that of the GSC's

labor. You can also ask your EAC to waive the supervisory fee, its minimum-hours requirement, and reduce the charges for supplies and materials (many of which you can provide at a lower cost).

Ⅱ. Please translate the following English sentences into Chinese.

1. 在经济不景气的情况下开展业务的好处之一是，会展管理层和行业供应商确实希望并需要您的业务。

2. 以下列表代表了您可以在整个贸易展管理过程中与之谈判的各种人员和供应商。

3. 许多参展商将其与展会管理的互动局限于后勤事务，如购买展位空间。

4. 许多展馆根据从地面到天花板的立方英尺①数量收取存储费，而不管您存储的物品是否实际占用了所有可用空间。

5. 我能够通过回收客户展台的铝制品来免除一家展馆所收取的展品处理费。

6. 回收收入抵消了剩余构筑物的填埋费。

7. 在人才公司，除了规定的人才日费率外，我还可以通过谈判免除他们通常收取的10%的代理费。

8. 如果您提出合理要求，并与供应商建立牢固的合作关系，他们将与您合作，以实现您的目标和预算。

Ⅲ. There are 10 sentences in this section. Beneath each sentence there are four words or phrases marked A, B, C and D. Choose one word or phrase that best completes the sentence.

1. C

2. B

3. D

4. A

5. D

6. C

7. A

8. D

9. C

10. D

Ⅳ. Writing

(略)

UNIT 9

展会知识

1. 参展商通常都是服务类型的公司，包括提供数字技术服务、生活服务、助农服务、供应链服务等。服贸会是服务交易，是非商品的交易，而广交会主体是商品交易。

2. 服务贸易是一国的法人或自然人在其境内或进入他国境内向外国的法人或自然人提供服务的贸易行为，包括商业服务、通信服务、建筑及有关工程服务、销售服务、教育服务、环境服务、金融服务、健康与社会服务等。

---

① 1立方英尺≈0.028 3立方米。

3. 跨境电子商务是指分属不同关境的交易主体，通过电子商务平台达成交易、进行支付结算，并通过跨境物流送达商品、完成交易的电子商务平台和在线交易平台。

B2B 平台：阿里巴巴国际站、中国制造网、环球资源、敦煌网等。

B2C 平台：速卖通、Wish、Shopee、Lazada、亚马逊等。

4. 跨境电子商务作为外贸新业态，对我国外贸的创新发展有着引领作用，也对我国经济发展有深远的影响。

（1）跨境电子商务作为"互联网+"国际贸易，能够突破时空限制，减少中间环节，解决供需双方信息不对称问题，为我国外贸提供发展新机遇，创造新的经济增长点。

（2）跨境电子商务是全球化时代的产物，是世界市场资源配置的重要载体。跨境电子商务平台将进一步打破全球市场壁垒，促进跨境商业流通，这必将促进中国的全面开放。

（3）跨境电子商务是推动产业结构升级的新动力，为企业打造国际品牌提供了新的机遇。

（4）跨境电子商务是消费时代的产物，它响应了国内消费者对更高生活质量的需求，能够进一步提高国内消费者的福利水平。

5. 开放性问题，学生自我思考。

📖 Reading 1

Ⅰ. Answer the following questions.

1. Linear booth, Perimeter booth, Peninsula booth, and Island booth.

2. (1) Theexhibitors have the opportunity to include marketing messaging on all sides.

(2) There are typically fewer restrictions around hanging signs compared to the other types of trade show booths.

(3) Theexhibitors also gain the flexibility to determine how attendees will come into, move about, and exit space.

(4) There's more breathing room between your booth and those of the other exhibitors.

3. Peninsula booths have aisles on three sides, and they share a wall with another exhibit. Peninsula booth tends to be smaller and less expensive than an island but more expensive than perimeter and linear booths. Dimensions are typically 20 feet by 20 feet or moderately larger.

4. (1) Be aware of whether your space is sharing a wall with an exhibit; if so, you'll need to know the rules and regulations on wall-sharing (like with a peninsula).

(2) Make use of the perimeter's unique dimensions—they may even influence your design concept.

(3) Confirm that the design you come up with works comfortably in the space.

5. Linear booth has the most limitations—limited space, limited accessibility, limited visibility, and so on. Plus, an enclosed space tends to feel smaller.

6. (1) The bottom of the canopy should not be lower than 7 ft. (2. 13 m) from the floor within 5 ft. (1. 52m) of any aisle. Canopy supports should be no wider than three inches 3 in. (0. 08 m).

(2) Fire and safety regulations in many facilities strictly govern the use of canopies, ceilings.

(3) Covered ceiling structures or enclosed rooms, including tents or canopies, shall have one smoke detector placed on the ceiling for every 900 square feet.

Ⅱ. Please translate the following English sentences into Chinese.

1. 与其他类型的展位相比，悬挂标识的限制通常较少。

2. 你可能没有足够的材料或信息填充四面，所以可以在其他空间投入更多。

3. 有时候缺少通道或者入口会破坏设计好的活动。

4. 你不用担心后墙，因为后面的空间无法通行。

5. 注意你的空间是否与其他展商共用一面展板（墙）。

6. 确认你提出的设计是否适用于这个空间。

7. 标准（道边）展位是展会中最常见的展位类型。

8. 因此，如果参加展会对你来说很重要但是你又要控制预算，标准（道边）展位是最好的选择。

9. 采用最全面的设计使空间得到最佳利用。

10. 但是，不同的会展中心，甚至展厅内部的规定都有所不同。

Ⅲ. There are 8 sentences in this section. Beneath each sentence there are four words or phrases marked A, B, C and D. Choose one word or phrase that best completes the sentence.

1. A

2. D

3. B

4. D

5. C

6. C

7. C

8. D

UNIT 10

展会知识

1. 世界博览会分为两种形式，一种是综合性世博会，另一种是专业性世博会。世博会是一项由主办国政府组织或政府委托有关部门举办的有较大影响和悠久历史的国际性博览活动。参展者向世界各国展示当代的文化、科技和产业上正面影响各种生活范畴的成果。

与传统意义上的商品展相比，世博会是由一个国家的政府主办、多个国家和国际组织参加的国际性大型博览会，是世界最高级别的展览活动。世博会的定位不是经济活动，而是通过展现人类在某一个或多个领域取得的文明进步，展望人类未来前景并教育大众。世博会的宗旨是促进各国人民之间更好地相互了解和沟通，维护世界和平。

2. 迪拜世博会的主题是"连接思想，创造未来"。迪拜世博会一是通过展示来自不同国家的发明创造，二是将园区划分为三个主要区域：流动性、机遇以及可持续性，通过建筑创新，来实现主题。

3.（1）世博会与奥运会的宗旨是一致的，都是为了让各国人民能更好地了解和沟通，建立美好、和平的世界。

（2）世博会与奥运会都是由国家政府主办，多个国家参与的大型国际活动。

（3）世博会和奥运会都是各国展示先进水平的平台，世博会是文明文化的展示，而奥运会是体育技能和精神的展示。

4. 学生自行查阅讨论。

5. 上海世界博览会于 2010 年 5 月 1 日至 10 月 31 日期间举行。

主题为：城市，让生活更美好（Better City, Better Life）。

上海世博会对中国经济发展的影响如下：

（1）为中国经济发展创造良好的外部环境。举办世博会可以提高我国的国际形象和地位，加强我国与其他各国的经济与技术合作，促进国际贸易的进一步发展。

（2）带动区域消费，促进消费需求的扩大。世博会的举办带动上海及周边地区旅游业、餐饮住宿、文化娱乐、服务业等的消费需求。

（3）为国内创造更多的就业机会。举办世博会带动了投资热潮，拉动区域经济发展，从而创造出更多的就业机会。

（4）世博会的后续经济效益强大。上海世博会的举办对于长三角地区的经济发展有显著效应，从而带动华东乃至全国的经济发展。

（可以围绕"世博经济"现象展开思考）

Reading 1

Ⅰ. Answer the following questions.

1. Globalization, digitization and new distribution forms are changing the consumer goods industry at a breakneck pace.

2. Digitization, networking, safety and security are some of the most pressing issues of our time.

3. The increasing scarcity of fossil fuels, growing global energy demand, access to clean water, waste disposal and recycling are all amongst the major challenges facing the world in future.

4. The industry's top themes are resource efficiency, production optimization, food safety, digital solutions and food trends.

5. Connected cars, autonomous driving and renewable energy drive systems are the hot-button topics in the sector.

6. With the ever-increasing importance of Industry 4.0, digitization, safety, security, textile care and cleanrooms have become essential components.

Ⅱ. Please translate the following English sentences into Chinese.

1. 我们的活动达到了最高的质量标准，并且在工业、商业、政治、服务和消费品之间建立了全球连接。

2. 我们在 28 个地点拥有约 2 200 名员工，几乎昼夜不停地在全球各地工作，以提升客户的利益。

3. 我们充当了供求之间、趋势和市场之间的桥梁。

4. 作为可靠的市场营销伙伴，我们是成效、质量和信任的代名词。

5. 这些挑战也为全球市场创造了新的机遇。

6. 该行业的主题是资源效率、生产优化、食品安全、数字解决方案和食品趋势。

7. 我们为整个纺织价值链创造动力。

8. 法兰克福展览公司将未来趋势与新技术、人与市场、供应与需求结合在一起。

9. 法兰克福拥有 30 多家博物馆，是欧洲博物馆种类最多的城市之一。

10. 法兰克福是欧洲大陆主要的金融中心之一。

Ⅲ. There are 10 sentences in this section. Beneath each sentence there are four words or phrases marked A, B, C and D. Choose one word or phrase that best completes the sentence.

1. D

2. D

3. D

4. C

5. D

6. B

7. B

8. C

9. B

10. B

Ⅳ. Writing

(略)